INTERNATIONAL DEVELOPMENT IN FOCUS

Financing Solutions to Reduce Natural Gas Flaring and Methane Emissions

GIANNI LORENZATO, SILVANO TORDO, BEREND VAN DEN BERG, HUW MARTYN HOWELLS, AND SEBASTIAN SARMIENTO-SAHER

WORLD BANK GROUP

Contents

Boxes

Figures

Map

Photos

Tables

Foreword

Decarbonizing the world's energy systems is an essential pillar of the global response to climate change. At present, gas flaring—the 160-year-old practice of burning the natural gas associated with oil extraction—and the related methane emissions represent as much as 12 percent of the greenhouse gases released by the global energy sector.

With the share of energy produced by oil and gas projected to increase until 2040, urgent action must be taken to accelerate the transition to net-zero. Ending routine gas flaring and curtailing methane emissions is an achievable way to significantly reduce emissions. This would not only help address climate change, it could also provide millions of people with an energy source. For instance, if all the associated gas flared across the world were put to productive purposes, it could power all of Sub-Saharan Africa.

Much attention has rightly been focused on large-scale gas capture projects that tackle the largest flares over the past two decades. At the same time, rapid technological innovation has made small-scale flaring reduction projects increasingly feasible. Correspondingly, total flaring volumes have decreased 19 percent since 2003.

However, these developments fail to address what we call the "missing middle": those flares where major financial investment is needed, but to which traditional financing approaches and solutions are ill-suited. Satellite-based estimates provided by the Global Gas Flaring Reduction Partnership (GGFR) show that these missing middle sites account for nearly 60 percent of the 142 billion cubic meters of associated gas burnt wastefully each year. These sites must now be prioritized.

This report provides a systematic framework for evaluating the feasibility of flare reduction projects at these locations. The approaches and tools developed will help policy makers and operators analyze investment barriers, identify key variables and success factors, and model financial options for those medium-size flares that have been overlooked historically.

Encouragingly, our findings show that project developers could attract the necessary investment from private investors and cost-effectively tackle many of the more than 2,000 flaring sites that fall into the missing middle. However, the report also finds that significant barriers remain for flaring and methane reduction projects, from a lack of prioritization by operators and unsupportive regulatory environments to inadequate infrastructure and macroeconomic risk. Overcoming these barriers requires creative, new ways of approaching flaring and methane reduction projects.

We believe this report helps signal the way forward and will encourage operators and governments to treat associated gas as an asset, not an unwanted by-product, enabling oil producers to end routine flaring by 2030.

Zubin Bamji
Program Manager
Global Gas Flaring Reduction Partnership
The World Bank

Acknowledgments

This report was commissioned and funded by the World Bank's Global Gas Flaring Reduction Partnership (GGFR), a multi-donor trust fund composed of governments, oil companies, and multilateral organizations working to end routine gas flaring at oil production sites across the world. GGFR helps identify solutions to the many technical and regulatory barriers to flaring reduction by developing country-specific flaring reduction programs, conducting research, sharing best practices, raising awareness, increasing the global commitments to end routine flaring, and advancing flare measurements and reporting.

The report was authored by Gianni Lorenzato (investment banking expert, consultant), Silvana Tordo (lead energy economist), Berend van den Berg (project developer), Huw Martyn Howells (senior consultant, GGFR), and Sebastian Sarmiento-Saher (research analyst).

Special thanks go to peer reviewers Pierre Audinet (lead energy specialist, World Bank), Ignacio de Calonje (chief investment officer, International Finance Corporation), and Babak Fayyaz-Najafi (senior planning engineer and company representative on the GGFR Steering Committee, Chevron) for helpful comments.

The authors thank Alex de Valukhoff, Dmitry Russkin, Billy Morley, Ilya Nesterenko, Pavel Yakushev, and Waleed Kamhieh at Aggreko plc; Gabriel Lorenzi and Octavio Molmenti at Galileo Technologies S.A.; Chris Levell at Gas Strategies Group Ltd; Franz Gruber at Hoerbiger Service, Inc.; Pablo Tribin at Mechero Energy; and Sumeet Singh and Pulak Sen at PowerGas LLC for their cooperation and supporting material for the case studies of this report. The authors also thank Charles Nyirahuku at the African Development Bank; Yo Takatsuki at AXA Investment Managers; Stephanie Saunier at Carbon Limits; Mark Davis and Brian Hepp at Capterio; Mounir Bouaziz, an independent consultant; Angel Marces at the Inter-American Development Bank; Michelle Horsfield at Climate Bonds Initiative; Tim Gould, Becky Schulz, KC Michaels, and Christophe McGlade at the International Energy Agency; Berit Lindholdt-Lauridsen, Aaron Levine, Francisco Avendano, and Ignacio de Calonje at the International Finance Corporation; James Mackey and Daniel Palmer at the Oil and Gas Climate Initiative; and Brigitte Bichler and Ernst Fietz at OMV Group for their insights and feedback.

The report was edited by Honora Mara. The authors would like to thank Clare Murphy-McGreevey (external affairs officer) for editorial guidance.

Executive Summary

This study aims to create awareness of the business case for reducing gas flaring and methane emissions, complementing the findings and recommendations of an upcoming World Bank study on the regulatory aspects of flaring (GGFR, forthcoming). It provides a framework for policy makers to evaluate the feasibility and financial attractiveness of flaring and methane emission reduction (FMR) projects, analyze investment barriers, identify key variables and success factors, and provide indicative financial modeling templates. The study also assesses the applicability of different sources of commercial capital to FMR financing and presents evidence and lessons learned from real case studies.

Gas flaring is a substantial source of carbon dioxide–equivalent emissions and a wasted opportunity to put gas to use. The World Bank's Global Gas Flaring Reduction Partnership (GGFR)[1] estimates that the total volume of natural gas flared globally was 142 billion cubic meters in 2020 (World Bank 2021). Although the volumes flared have decreased over time, the world still flares enough gas to power Sub-Saharan Africa. Emissions from flared gas and methane at oil and gas operations represent about 12 percent of global energy sector greenhouse gas emissions.

The monetization of associated gas can be an attractive complement to regulation and a financial solution to flaring. Robust regulation is a necessary but not always sufficient incentive for oil companies to reduce flaring. Regulation is crucial to creating an enabling and compulsory environment to eliminate routine gas flaring from oil production sites. Such an environment is particularly important in certain cases, for example, when dealing with a large number and volume of small flare sites, or in countries where the regulatory framework is nonexistent or not enforced, or situations in which gas flaring projects are uneconomic. Treating associated gas as an asset, not an unwanted by-product of oil production, offers a path forward.

Various technologies are available to recover and monetize associated gas, instead of flaring it into the atmosphere. Associated gas is customarily reinjected into the field—the default solution to flaring in many oil fields. When reinjection is not possible or its capacity has been exhausted, six solutions can be considered, depending on the specific features of the field. Associated gas can be (1) converted to power, with the latter used on-site by the oil operator; (2) converted to power, with the latter sold to external off-takers; (3) delivered to an existing pipeline network; (4) delivered to a gas processing plant; (5) compressed and sold as compressed natural gas; or (6) liquefied for sale as liquefied natural gas.

The study focuses on FMR projects at midsized flares that are too small to be prioritized by oil companies but still amenable to profitable monetization ("missing middle" flares). According to GGFR, about 2,358 flare sites fall in the 1 million standard cubic feet per day (mmscf/d) to 10 mmscf/d range and represent 58 percent of global flare volumes. Below that range, FMR projects are unlikely to be economically viable, unless clustered in larger projects or propelled by an enabling and compulsory regulatory framework. Above that range, there are large and mega flares whose recovery requires highly tailored projects, large ancillary infrastructure (for instance, for gas or electricity transmission), government planning, and capital injections in the hundreds of millions of dollars.

Financial models developed for this study indicate that FMR projects in the missing middle can offer potentially attractive equity returns to project developers. Financial returns were modeled on the basis of indicative assumptions derived from project experience and feedback from industry experts. A base case scenario was defined for each of the six technologies at three different flare sizes (1 mmscf/d, 5 mmscf/d, and 10 mmscf/d). Although some of the assumptions may be too stylized to reflect the variety of contexts in which real FMR projects take place, these financial models are meant as user-friendly tools for policy makers to conduct a preliminary assessment of FMR project feasibility and potential attractiveness to private investors. The models indicate the following: FMR projects are attractive at 10 mmscf/d flares (double-digit internal rates of return); returns are still positive but sometimes in the single digits at 5 mmscf/d flares; and, at 1 mmscf/d flares, internal rates of return are negative or single-digit at best, making the clustering of small flares a necessity if feasible.

Several barriers, however, affect the financing and execution of FMR projects. Some barriers are common to all flare sizes, whereas others are specific or magnified in the case of missing middle flares:

- *Small FMR projects are seen as an unworthy diversion* of engineering and management resources by oil companies, because the economics of such projects are trivial compared to those of the core oil production business.

- *The amount, reliability, and quality of associated gas supply are never known with certainty up front.* Flare size and decline rate are subject to the same unpredictability as the oil production profile of a field. Mindful of this unpredictability, oil companies are unlikely to enter into associated gas deliver-or-pay contracts.

- *Access to existing infrastructure may be absent or inadequate.* Small and geographically dispersed flare sites may require prohibitive investment in ancillary infrastructure for FMR projects to take place (for example, power lines for gas-to-power projects, associated gas transport infrastructure).

- *The regulatory environment may not be supportive*, for instance, because of (1) absence, insignificance, or poor enforcement of flare fines; (2) unclear rules granting ownership of associated gas to the oil producer; and (3) lack of rules granting FMR developers access to flare sites and flare site data.

- *End-product (gas or power) prices may be too low or volatile.* For instance, associated gas-to-power projects may not be attractive in highly competitive wholesale electricity markets with low grid prices.

- *Off-taker payment risk.* Like other infrastructure projects, FMR projects are exposed to the risk that the off-taker (for example, power distribution company) may not honor its contractual obligations.

- *FMR projects are subject to a variety of macroeconomic and political risks,* including those affecting oil companies, because the supply of associated gas goes hand in hand with oil production. Project execution can also be complex and require coordination with several stakeholders (oil company, regulator, off-taker, local communities).

Navigating these barriers requires project developers with specific FMR expertise, as highlighted by the six detailed case studies compiled for this report, which are based on extensive interviews with and information shared by project executives. The case study selection (see chapter 4) attempts to reflect the technical and geographic diversity of FMR projects. From a technical standpoint, three case studies discuss gas-to-power projects (Aggreko, Hoerbiger, and Mechero), one discusses a liquefied natural gas project (Galileo), and one an innovative digital flare mitigation approach (Crusoe Energy). From a geographic standpoint, the case studies discuss projects and solutions applied in Latin America, the Middle East, Nigeria, North America, and the Russian Federation. Five case studies cover actual projects whereas one (Nigeria) describes a novel regulatory approach to FMR.

    The case studies and additional input from industry experts underscore the importance of several best practices to address FMR project barriers and achieve attractive financial returns. Although these best practices cannot address all barriers—especially those of an external nature, such as end-product price volatility or macro risk—they can increase the odds of project success. Specific best practices and lessons learned include the following:

- *The FMR developer's ability to provide turnkey solutions is key.* Oil companies are more likely to pull the trigger on FMR projects if FMR developers are able to take care of the whole process—including design, procurement of equipment (or use of developer-owned equipment), installation, financing, and operation and maintenance.

- *Modular and movable equipment is critical to respond to changes in flare profile.* FMR developers that deploy modular equipment can easily down- or upsize their operations during the course of a project and shift equipment among different flares within a reasonable geographic range.

- *A portfolio approach that clusters several small flares under the same project* is often required to build a minimum of economies of scale and hedge against the uncertainty and unpredictability of flare profiles.

- *The FMR developer's ability to equity-finance a project is critical to achieve financial closure.* Because of the risks discussed previously, many FMR projects do not meet the leverage requirements of traditional nonrecourse project finance. If available, leverage may be lower than in typical infrastructure projects. Developers must be ready to fund construction through own equity or to mobilize other equity sources.

- *Strong project management and execution capabilities are essential.* Execution risk is heightened in the FMR arena by the variety of stakeholders involved (oil company, off-taker, regulator, and local communities, among others) and the geographic dispersion of flares within a site.

The report also identifies various sources of capital that already or could potentially finance FMR projects, with various degrees of applicability depending on project size and complexity:

- A number of project developers have expertise specific to FMR and have the ability to not only design, implement, and operate projects but also provide all or part of the equity required.

- Similarly, equipment suppliers to the oil and gas industry—especially those focused on gas processing and compression technologies and power generation units—have identified FMR as an interesting niche and are taking on the role of project developers and investors. These suppliers increasingly install, operate, and maintain equipment at their expense, under long-term contracts.

- Strategic investment funds, particularly those sponsored and funded by fossil fuel–rich countries that are looking to decarbonize their economies, are well placed to support FMR projects given their high degree of local knowledge and visibility over project pipelines.

- Commercial banks represent a large pool of capital for the oil and gas sector, despite pressure to reduce exposure for environmental, social, and corporate governance reasons. FMR lending could be considered a step toward greening of the banks' oil and gas exposures. Small project size, however, may be an issue, especially for international banks.

- Private capital funds—especially those with expertise in the energy sector—are also a potential source of funding for FMR projects, particularly as equity providers.

For all categories of private capital providers, the transition toward a low-carbon future is becoming an ever more important factor driving sustainable investment and capital allocation. This priority should create an incentive to consider FMR investments, which could also be eligible for the issuance of new securities such as transition bonds and sustainability-linked loans.

At the same time, increasingly stringent criteria apply to the financing of FMR projects by development finance institutions, and their classification as mitigation finance is now restricted to brownfield projects that substantially reduce greenhouse gas emissions. Under prevailing international standards, the oil industry is not eligible to issue green bonds and loans. But FMR projects may be eligible for the new asset class of transition bonds and loans, which is still a small market niche. In addition, there is no evidence to date that FMR projects qualify for the sale of carbon credits in mandatory emissions trading schemes or voluntary carbon credit markets, which could improve FMR returns and attract more private capital to the sector. Regulation imposing strict flare payments on oil companies, however, could achieve an effect similar to the sale of carbon credits if the oil companies contractually agree to pass some of the savings from reduced flare payments on to the FMR project developers.

NOTE

1. For more information on the GGFR, see https://www.worldbank.org/en/programs/gasflaringreduction#7.

REFERENCES

GGFR (Global Gas Flaring Reduction Partnership). Forthcoming. *Global Review of Regulation of Gas Flaring and Venting.* Washington, DC: World Bank.

World Bank. 2021. "Seven Countries Account for Two-Thirds of Global Gas Flaring." Press Release 2021/143/EEX, April 28. https://www.worldbank.org/en/news/press-release/2021/04/28/seven-countries-account-for-two-thirds-of-global-gas-flaring.

Abbreviations

AG	associated gas
bcm	billion cubic meters
Btu	British thermal unit
capex	capital expenditures
CBI	Climate Bonds Initiative
CNG	compressed natural gas
CO_2	carbon dioxide
DFI	development finance institution
DFM	digital flare mitigation
DPR	Department of Petroleum Resources (Nigeria)
EBRD	European Bank for Reconstruction and Development
EPC	engineering, procurement, and contracting
ESG	environmental, social, and corporate governance
EU	European Union
EU GBS	European Union Green Bond Standard
FMR	flaring and methane reduction
GBP	Green Bond Principles
GGFR	Global Gas Flaring Reduction Partnership
GHG	greenhouse gas
GPP	gas processing plant
$GtCO_2e$	gigatons of carbon dioxide equivalent
H_2S	hydrogen sulfide
IEA	International Energy Agency
IFC	International Finance Corporation (of the World Bank Group)
IOCs	international oil companies
IRR	internal rate of return
km	kilometer
kV	kilovolt
kWh	kilowatt-hour
LNG	liquefied natural gas
LPG	liquefied petroleum gas
m^3	cubic meter
mmBtu	million British thermal units

mmscf/d	million standard cubic feet per day
mscf	thousand standard cubic feet
Mt	million tons
$MtCO_2e$	million tons of carbon dioxide equivalent
MW	megawatt
NDC	Nationally Determined Contribution
NGFCP	Nigerian Gas Flare Commercialisation Programme
NGL	natural gas liquid
NOAA	National Oceanic and Atmospheric Administration (United States)
NPV	net present value
O&M	operations and maintenance
OCM	original component manufacturer
opex	operating expenditures
psi	pounds per square inch
SIF	strategic investment fund
SLL	sustainability-linked loan
tCO_2e	tons of carbon dioxide equivalent
TMM	Termo Mechero Morro
VIIRS	Visual Infrared Imaging Radiometer Suite

1 Gas Flaring and Methane Emissions Facts and Trends

QUANTIFYING GREENHOUSE GAS EMISSIONS FROM OIL AND GAS OPERATIONS

In 2015 oil and gas extraction, processing, and transportation ("oil and gas operations") were responsible for an estimated 4.4 gigatons of carbon dioxide equivalent ($GtCO_2e$) emissions, representing 9 percent of all human-made greenhouse-gas (GHG) emissions (Beck et al. 2020; Olivier and Peters 2018).[1] According to the International Energy Agency (IEA), by 2019 this estimate had already risen to 5.4 $GtCO_2e$—approximately 15 percent of global energy sector GHG emissions, partly on account of better measurement (IEA 2020d, e). Over half of these emissions (2.7 $GtCO_2e$) came from flaring and methane released during oil and gas operations. These estimates include *scope 1* and *scope 2* emissions. The former are direct emissions from oil and gas operations, whereas the latter are indirect emissions arising from the generation of energy purchased by the oil and gas industry as part of its activities (box 1.1).

An additional 16.2 $GtCO_2e$ emissions, representing 33 percent of all human-made GHG emissions in 2015, occurred during the combustion of fuel by end users, for instance car drivers (Beck et al. 2020; Olivier and Peters 2018).[2] These are known as *scope 3* emissions (figure 1.1). Scope 3 emissions are highly dependent on consumer choices and their attribution to the oil and gas industry is debatable.[3] For those reasons, scope 3 emissions are beyond the scope of this study.

Global oil and gas emissions fell in 2020 as a result of the decline in global demand associated with the COVID-19 (Coronavirus) pandemic (IEA 2021a). However, the IEA (2021a) points at evidence of rapidly increasing energy demand as the world gradually recovers from the COVID-19 pandemic, suggesting that CO_2 emissions may rebound to precrisis levels depending on governments' emphasis on clean energy transition in their efforts to reboot economic growth. Consequently, the analysis presented in this chapter is primarily based on 2019 emissions data, although 2020 references are also provided.

BOX 1.1

Scope 1, scope 2, and scope 3 emissions

Scope 1 emissions come directly from the oil and gas industry itself, such as emissions from powering the engines of drilling rigs, leaks of methane in the upstream or midstream phases of the production cycle, and emissions from ships used to transport oil or gas. They are also referred to as "direct emissions" from oil and gas operations.

Scope 2 emissions arise from the generation of energy purchased by the oil and gas industry, such as the generation of electricity taken from a grid to power auxiliary services, or the production of hydrogen purchased by a refinery from an external supplier. They are also referred to as "indirect emissions" from oil and gas operations.

Source: IEA 2020a.

The sum of scope 1 and scope 2 emissions is sometimes referred to as "well-to-tank" or "well-to-meter" emissions. The International Energy Agency estimates that 95 kilograms of carbon dioxide equivalent is emitted on average to bring a barrel of oil to end consumers, with significant variations. For natural gas, the comparable average is 100 kilograms of carbon dioxide equivalent per barrel of oil equivalent, also within a broad range.

Scope 3 emissions occur during combustion of fuel by end users, for instance car drivers. Scope 3 emissions vary substantially depending on the type of oil and gas products combusted.

FIGURE 1.1

Sources of greenhouse gas emissions (scope 1 and scope 2) from oil and gas operations, 2019

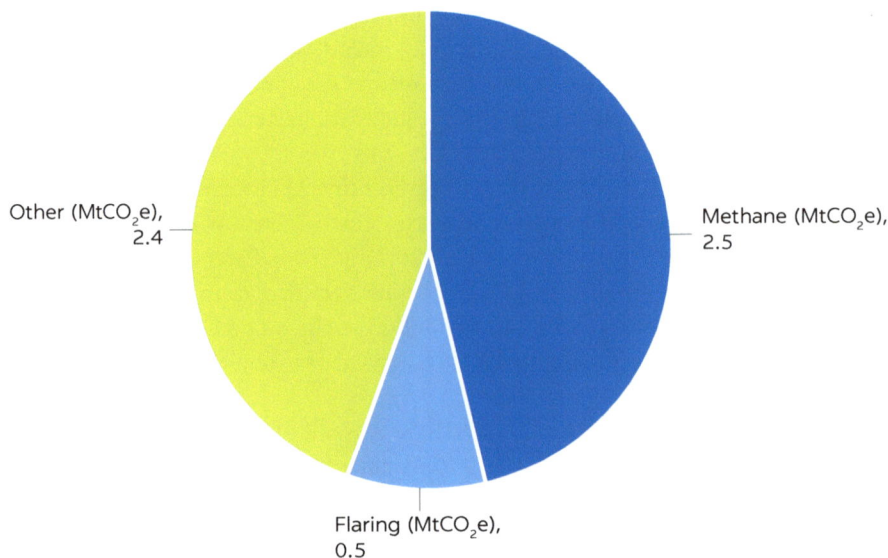

Other (MtCO$_2$e), 2.4

Methane (MtCO$_2$e), 2.5

Flaring (MtCO$_2$e), 0.5

Sources: World Bank estimates of 2019 oil and gas total emissions based on IEA 2020b.
Note: "Other" includes scope 2 (indirect) emissions as well as scope 1 (direct) emissions not due to flaring or methane emissions, such as emissions from operating oil and gas production facilities, powering drilling rigs, and ships used to transport oil or gas. MtCO$_2$e = million tons carbon dioxide equivalent.

Over half of scope 1 and scope 2 emissions from oil and gas operations come from two sources (figure 1.1):

1. *CO_2 emissions from the flaring of natural gas, particularly gas associated with oil production.* Natural gas is a normal by-product of oil production. Often, operators cannot find a productive use for it, such as use as a source of power or heat on-site for own consumption, reinjection into oil wells to create pressure for secondary liquids recovery, operating and infrastructure restrictions, or sale to end consumers. In these circumstances, this "associated gas" is generally flared (upstream flaring), resulting in the emission of CO_2. Gas is also flared downstream at liquefied natural gas plants and gas processing facilities when volumes produced exceed the capacity of gas infrastructure. In 2019, the emissions from upstream and downstream flaring accounted for 8.6 percent of scope 1 and scope 2 emissions from oil and gas operations, or some 462 million tons of CO_2 equivalent ($MtCO_2e$).[4]

2. *Methane (CH_4) emissions from the intentional or accidental release of natural gas in the atmosphere without ignition.* Methane emissions can be released at different phases of the oil and gas value chains, including production, collection, and processing of gas, and transmission and distribution to end consumers. Some methane emissions are accidental, for example, because of leaking valves (usually called "fugitive emissions"). Others are deliberate, often carried out for safety reasons, as part of the design features of a facility, or because of operating restrictions, and are usually called "vented emissions" (IEA 2020a). Some venting occurs also during gas flaring, because the latter process is rarely 100 percent efficient, and a fraction of the natural gas passes through unburned.[5] In this study, unless specified otherwise, *methane emissions* refer to the sum of fugitive and vented emissions. In 2019, these emissions represented 46.2 percent of scope 1 and scope 2 emissions from oil and gas operations—almost 82 Mt or about 2.5 $GtCO_2e$ (IEA 2020b).

Although CO_2e emissions from methane are far higher than those from flaring, methane emission volumes (estimated at 72 Mt, down from 82 Mt in 2019) are far smaller than flared volumes (estimated at 152 billion cubic meters [bcm] in 2020, down from 159 bcm in 2019).[6] This discrepancy exists because, although methane has a much shorter atmospheric lifetime than CO_2 (about 12 years, compared with centuries for CO_2), it is a much more potent GHG. The Intergovernmental Panel on Climate Change estimates a global warming potential for methane of 84–87 tons of CO_2 for each ton of methane when considering its impact over 20 years and 28–36 tons of CO_2 over 100 years (IEA 2020a).

From a value chain standpoint, CO_2 emissions are concentrated in the upstream oil production phase, whereas methane emissions occur more broadly across the value chain. Figure 1.2 shows the estimated breakdown of direct and indirect oil and gas emissions across the value chain according to Beck et al. (2020).

FIGURE 1.2

FIGURE 1.2

Direct and indirect emissions from oil and gas operations, by source, 2015

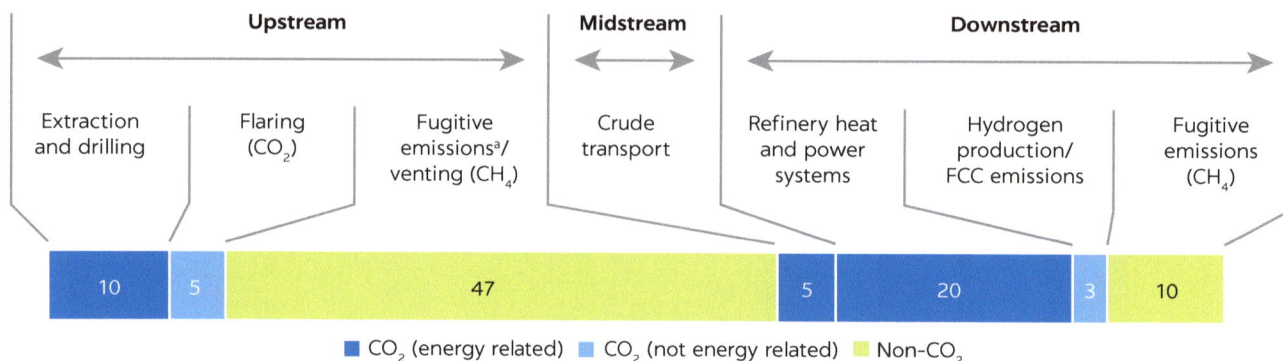

| Upstream | Midstream | Downstream |

| Extraction and drilling | Flaring (CO_2) | Fugitive emissions[a]/ venting (CH_4) | Crude transport | Refinery heat and power systems | Hydrogen production/ FCC emissions | Fugitive emissions (CH_4) |

| 10 | 5 | 47 | 5 | 20 | 3 | 10 |

■ CO_2 (energy related) ■ CO_2 (not energy related) ■ Non-CO_2

Source: Based on Beck et al. 2020.
Note: CH_4 = methane; CO_2 = carbon dioxide; FCC = fluid catalytic converter.
a. Fugitive emissions from midstream are included in upstream (about 20 percent of total oil and gas emissions, mainly methane) to be consistent with IEA (2018) classification.

GAS FLARING

Key concepts

Flaring is the controlled burning of gas during oil and gas operations. Under ideal conditions, a well-designed and well-operated flare typically has a combustion efficiency of 98 percent—that is, less than 2 percent of the gas passes through the flare stack unburned. Field studies, however, indicate that combustion efficiency is often several percentage points lower. Gas flared transforms into CO_2, and the amount unburned is vented in its original form as methane and usually some heavier hydrocarbons. The unburned methane, with its high global warming potential, significantly increases the GHG emissions from the flaring process. For example, if a flare burns at an efficiency of just 95 percent (that is, 5 percent of the gas passes through the flare stack unburned), the previously mentioned emissions from flaring in 2020 (440 $MtCO_2e$) increase by 11 percent to 488 $MtCO_2e$; with a 90 percent flare efficiency, the emissions increase by 29 percent to 568 $MtCO_2e$. Factors such as gas composition, flare stack design, flow rates, gas exit velocities, and steam use contribute to the combustion efficiency, which makes measuring the combustion efficiency of a flare on a continuous basis a technically complex endeavor (Ipieca 2018).

Associated gas is natural gas obtained as a by-product of oil extraction. Most oil wells also yield a mixture of other hydrocarbons such as condensates, natural gas liquids, and natural gas—the last is referred to as "associated gas." Oil production from a well is usually sent to a separator, where it is divided into its base components of oil, gas, and water. The oil is processed to remove impurities and transported to a refinery (via pipeline or tanker). The extracted gas is typically (1) used as fuel to generate electrical or mechanical power for own consumption; (2) reinjected into the well to maintain reservoir pressure; or (3) processed, compressed, and transported via pipeline for sale (Argonne Venting and Flaring Research Team 2017). When none of these approaches is operationally or economically attractive, associated gas is usually flared.

Flaring of associated gas in oil and gas production sites represents over 90 percent of global flaring. Flaring occurs in three sectors: (1) upstream flaring of associated gas during oil production; (2) mid/downstream at gas processing plants, liquefied natural gas terminals, and oil refineries, and in the petrochemical industry; and (3) in the industrial sector (for example, coal mines, landfills, cement production, and water treatment plants).[7] The Global Gas Flaring Reduction Partnership (GGFR), in partnership with the US National Oceanic and Atmospheric Administration (NOAA) and the Colorado School of Mines, has developed global gas flaring estimates based on observations from a satellite launched in 2012 and equipped with a Visual Infrared Imaging Radiometer Suite (VIIRS) of detectors. In 2020, VIIRS automatically detected about 10,000 active flares around the globe. Using VIIRS data, GGFR estimates that the total volume of natural gas flared globally in 2020 was 152 bcm (some 7 bcm less than in 2019, mainly as result of lower demand during the COVID-19 pandemic), of which upstream flaring accounted for 91 percent and mid/downstream for 9 percent.[8] Exact estimates of flared gas volumes by sector (oil vs. gas) and phase of the value chain (upstream vs. mid/downstream) are not available. Using input from GGFR, table 1.1 attempts such mapping on a purely indicative basis.

Depending on the rationale for its application, flaring can be categorized as routine, safety, and nonroutine (GGFR 2016). Table 1.2 contains examples of the following types of flaring:

- *Routine flaring* is undertaken during normal oil production operations in the absence of sufficient facilities or amenable geology to reinject the produced gas, use it on-site, or dispatch it to a market.[9] For instance, an upstream operator may flare because the sale of associated gas to end consumers is hindered by the remoteness and topology of the drilling location, low gas price in accessible markets (discouraging an investment in gas transportation infrastructure), or a time lag between the development of the field and connection to a gas pipeline (IEA 2020a). The reasons for routine flaring are explained further in the next subsection of this chapter.

- *Safety flaring* is done to purge flare lines, to ensure safe operation of the facility, or to do both, and includes emergency shutdowns.

- *Nonroutine flaring* can occur for a range of reasons outside of routine and safety, such as equipment maintenance, initial field start-up or start-up after a shutdown, temporary failure at gas collection facilities, and well testing and maintenance.

TABLE 1.1 **Volume of gas flared globally, by sector and phase, 2020**

SECTOR	UPSTREAM FLARING	MID/DOWNSTREAM FLARING
Oil sector	~ 135 bcm	~ 10 bcm
	Mostly associated gas flaring at oil production sites	Gas from refineries and petrochemical plants
Gas sector	~ 4 bcm	~ 3 bcm
	Mostly gas from condensate stripping in northern Russian gas fields	Gas from LNG liquefaction and gas processing plants
Total volume flared	~ 139 bcm	~ 13 bcm

Source: Global Gas Flaring Reduction Partnership data (https://www.worldbank.org/en/programs/gasflaringreduction#7).
Note: Definitions of upstream, midstream, and downstream vary across companies and research institutions. In this study, upstream is defined as activities involving the exploration, development, and production of oil and gas, consistent with Ipieca (2018). Midstream and downstream are consolidated in a single category, encompassing the transportation of crude oil to refineries and natural gas to processing facilities, and the refining, marketing, and distribution of oil and gas products to end users. Under this convention, liquefied natural gas and gas-to-liquids activities belong to the mid/downstream phase. bcm = billion cubic meters; LNG = liquefied natural gas.

TABLE 1.2 **Types of gas flaring**

TYPE OF FLARING	EXAMPLES
Routine	• Flaring from oil/gas separators • Flaring of gas production that exceeds existing gas infrastructure capacity • Flaring from process units such as oil storage tanks, tail gas treatment units, glycol dehydration facilities, and produced water treatment facilities, except where required for safety reasons
Safety	• Gas stemming from an accident or incident that jeopardizes the safe operation of the facility • Blow-down gas following emergency shutdown to prevent overpressurization of all or part of the process system • Gas required to maintain the flare system in a safe and ready condition (purge gas / make-up gas / fuel gas) • Gas required for a flare's pilot flame • Gas produced as a result of specific safety-related operations, such as safety testing, leak testing, or emergency shutdown testing • Gas containing H_2S, including the volume of gas added to ensure good dispersion and combustion • Gas containing high levels of volatile organic compounds other than methane
Nonroutine	• Flaring as a function of irregular (pronounced peaks and lows) gas production profile • Temporary (partial) failure of equipment that handles the gas during normal operations, until their repair or replacement (for example, failure of compressors, pipeline, instrumentation, or controls) • Temporary failure of a customer's facilities that prevents receipt of the gas • Initial plant/field start-up before the process reaches steady operating conditions or before gas compressors are commissioned • Start-up following facility shutdowns • Scheduled preventive maintenance and inspections • Construction activities, such as tie-ins, change of operating conditions, and plant design modifications • Process upsets when process parameters fall outside the allowable operating or design limits and flaring is required to restabilize the process • Reservoir or well maintenance activities such as acidification, wire line interventions • Exploration-, appraisal-, or production-well testing or cleanup following drilling or well work-over

Source: GGFR 2016.
Note: H_2S = hydrogen sulfide.

Despite the lack of exact measurements, GGFR believes that up to 85 percent of current total flaring is routine. Routine flares are more prone to recovery models involving gas commercialization, whereas nonroutine flares are better tackled through operational improvements.

Routine flaring—reasons and technical solutions

In the upstream phase, oil and gas operators routinely flare associated gas for a combination of operational, economic, and regulatory reasons, as summarized in table 1.3.

Reductions in gas flaring and vented gas are attractive options to reduce GHG emissions because of the high concentration of emissions per point source and the fact that gas is a marketable commodity, which reduces the mitigation costs per ton of CO_2. Use of gas that would otherwise be completely lost displaces the need for new sources of GHG-emitting fossil fuels (Elvidge et al. 2018). In practice, a combination of the operational, economic, and regulatory constraints determines under which conditions a technology represents a worthwhile investment. A range of technologies is available to recover and monetize gas routinely flared, and deliberately vented gas (see table 1.4). Chapter 3 presents a detailed analysis of the conditions required for the implementation of these technologies, as well as an assessment of which technologies are likely to have the broadest application in flared and vented gas recovery.

TABLE 1.3 Reasons for routine flaring and venting (upstream)

CATEGORY	REASONS
Operational reasons	• Lack of adequate gas gathering, transport, and processing infrastructure close to the extraction area • Remote locations of oil fields and offshore platform that make it technically impossible to build pipelines or electricity lines to use associated gas (rarely the case) • Infeasibility or inefficiency of reinjecting gas in underground reservoirs (depending on geology) • Well testing/completions cleanup in shale-oil/shale-gas developments
Economic reasons	• Flaring or venting of gas that cannot be sold commercially or stored underground • Remote locations of oil fields and offshore platform that make it uneconomic to build pipelines or other infrastructure to monetize associated gas • Competition from cheaper, reliable long-term supply of nonassociated gas • Local gas market (if present) that is too small to justify the investment • Low gas prices in potentially accessible market(s) • Market barriers to the transportation and sale of gas (for example, presence of a local monopolist) • Lack of economy of scale to justify investment in gas recovery infrastructure due to small-scale scattered flares or venting sites, or steep production decline curve (common for mature fields) • Associated gas forecast and production profile • Financial barriers to implementing infrastructure investments to gather, transport, and process gas • Unattractive cost-benefit analysis of removing sulphur and other contaminants from natural gas, pressurizing and transporting gas to customers, or removing liquid fractions before gas enters the pipeline system • Reinjecting gas in underground reservoirs not economical (depending on geology) • Country-specific risk (for example, political instability) that discourages investments
Regulatory reasons	• Lack of a robust and transparent data management system • Absence of effective regulatory frameworks on flaring or venting • Absence of powerful enforcement authorities • Limited resources hindering the ability of regulators to monitor levels and conditions of flaring and venting • Lack of government incentives to minimize flaring and venting • Lack of clear ownership of associated gas in some production-sharing agreements • Preemptive rights of host governments to purchase associated gas at unattractive conditions (for example, without contributing to the necessary infrastructure investment)

Source: Based on Buzcu-Guven and Harriss 2012.

Global gas flaring volumes

In 2019, the combined estimated volume of gas flared (159 bcm) and vented into the atmosphere (55 bcm) was almost equal to the volume of natural gas imports of China and Japan combined, or about 5 percent of total natural gas production.[10] In the same year, the portion of associated gas flared, vented, and fugitive ranged from about 30 percent in North America to 46 percent in Africa (figure 1.3). Gas that was not flared or vented was used on-site as a source of power or heat, reinjected into oil wells to create pressure for secondary liquids recovery, or sold to end consumers via gas grids.[11]

Hypothetically, if all the gas flared could be used for electricity production, it would be sufficient to generate over 700 billion kilowatt-hours of electricity, enough to power the entire African continent.[12] Valued at a wholesale price of US$2.00 per million British thermal unit, the gas flared in 2019 would be worth about US$10.6 billion and that in 2020 worth about US$10.0 billion. Although these estimates are extreme simplifications—for instance, they disregard the wide range of natural gas prices in different regions and the actual feasibility of selling flared gas—they provide a sense of the magnitude of the issue. In addition, the recovery of flared gas could have positive development impact in locations where communities lack reliable energy supply or rely on biomass as fuel.

TABLE 1.4 **Technologies for the recovery of flared or vented gas in the upstream phase (excluding fugitive emissions)**

TECHNOLOGY	STATUS	SUMMARY DESCRIPTION
Reinjection	Mature	Gas compression to repressurize areas of oil reservoirs, to maintain/increase oil production, or for underground storage. Not applicable in all geological formations.
Liquefied petroleum gas (LPG)	Mature	Gas compression and cooling to separate the heavy carbon fraction of the gas to produce LPG. Associated gas has a higher percentage of propane and butane compared to nonassociated gas. Storage and transportation of LPG to local markets is generally relatively easy.
Liquefied natural gas (LNG)	Mature	Technology to transform natural gas into liquid by cooling down to approximately negative 162°C (negative 259°F), and to store before shipment via LNG carriers (offshore) or containers (onshore). Onshore transport of gas as LNG is often called a "virtual pipeline."
Floating LNG	Under development for small-scale applications	New technology combining floating deepwater offshore production with LNG production (not generally applicable for small-scale gas volumes).
Compressed natural gas (CNG)	Mature	Technology to compress natural gas to 1/200th or less of the original volume, store, and transport in cylinders under pressure. Potentially a cost-effective alternative to LNG for small volumes / short distances to market (less than ~500 km). Transport of gas as CNG is often called a "virtual pipeline."
Natural gas hydrate (NGH)	Very early stage	Technology to crystallize natural gas in solid form. NGH is less dense than LNG but cheaper to produce, store, and transport. Applications to associated gas are being researched.
Adsorbed natural gas (ANG)	Early stage	Storage of natural gas in container with high adsorption capacity for gas (for example, metal-organic frameworks—MOFs), opening the potential to store more gas than in other vessels.
Gas-to-liquids (GTL)	Mature (early stage for small-scale applications)	Technology to chemically convert gas into synthetic crude oil, transportation fuels, naphtha, and other specialty liquids.
Gas conversion to methanol, dimethyl ether, and ammonia	Mature (early stage for small-scale applications)	Technology to convert methane found in associated gas to methanol (and subsequently dimethyl ether) or ammonia. Commonly used in Persian Gulf oil-producing states.
Gas-to-power	Mature	Construction of gas-fired power plants to supply electricity for on-site field operations or transportation to consumers via electricity lines.

Sources: Based on Argonne Venting and Flaring Research Team 2017; Buzcu-Guven and Harriss 2012; GGFR 2019b; Hajilary, Rezakazemi, and Shahi 2020.

FIGURE 1.3

Use of associated gas, by region, 2019

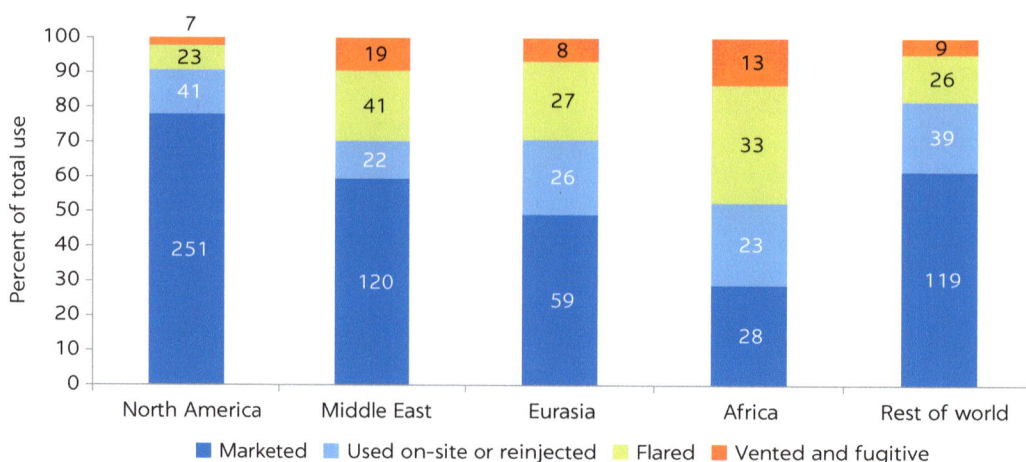

Source: Schulz, McGlade, and Zeniewski 2020.
Note: Values in bars for each region show billion cubic centimeters of use, by type of gas.

About 54 percent of the gas volume flared in 2020 was concentrated in 5 countries (in order of volume flared: the Russian Federation, Iraq, the Islamic Republic of Iran, the United States, and Algeria) and 73 percent in 10 countries. Russia is the largest gas flaring country, at approximately 25 bcm in 2020, closely followed by Iraq and the Islamic Republic of Iran, at about 17 and 13 bcm respectively, and the United States, at 12 bcm (figure 1.4).

The volume of gas flared globally decreased by 14 percent from 1996 to 2020, whereas oil production increased by 19 percent over the same period (figure 1.5). This diverging trend may reflect greater efficiency in associated gas use, tighter regulation and enforcement, or a lower gas-to-oil ratio in recently developed fields—factors that may have contributed to the steady decrease in gas flaring intensity. The volume of gas flared in a country primarily reflects a combination of (1) the size of a country's oil and gas industry, (2) the volume of gas produced per barrel of oil (the gas-to-oil ratio), and (3) the volume of gas used or reinjected.

Flaring volumes are highly concentrated by oil field, with the 20 largest flaring fields representing nearly 20 percent of global volumes in 2020. Most flare sites are between 0.25 and 10.00 million standard cubic feet per day (see chapter 3, figure 3.3).

A proxy measure of the effectiveness of a country's gas use is flaring intensity, measured in cubic meters of gas flared per barrel of oil produced.[13] High flaring volumes in Russia, the United States, and the Persian Gulf states reflect primarily the sheer size of the oil and gas industry in those countries; but their flaring intensity is relatively low (figure 1.6). Globally, flaring intensity in 2020 was 5.1 cubic meters, up from 5.0 cubic meters per barrel of oil produced in 2019.

FIGURE 1.4

Top 30 flaring countries, by 2020 volume, 2015–20

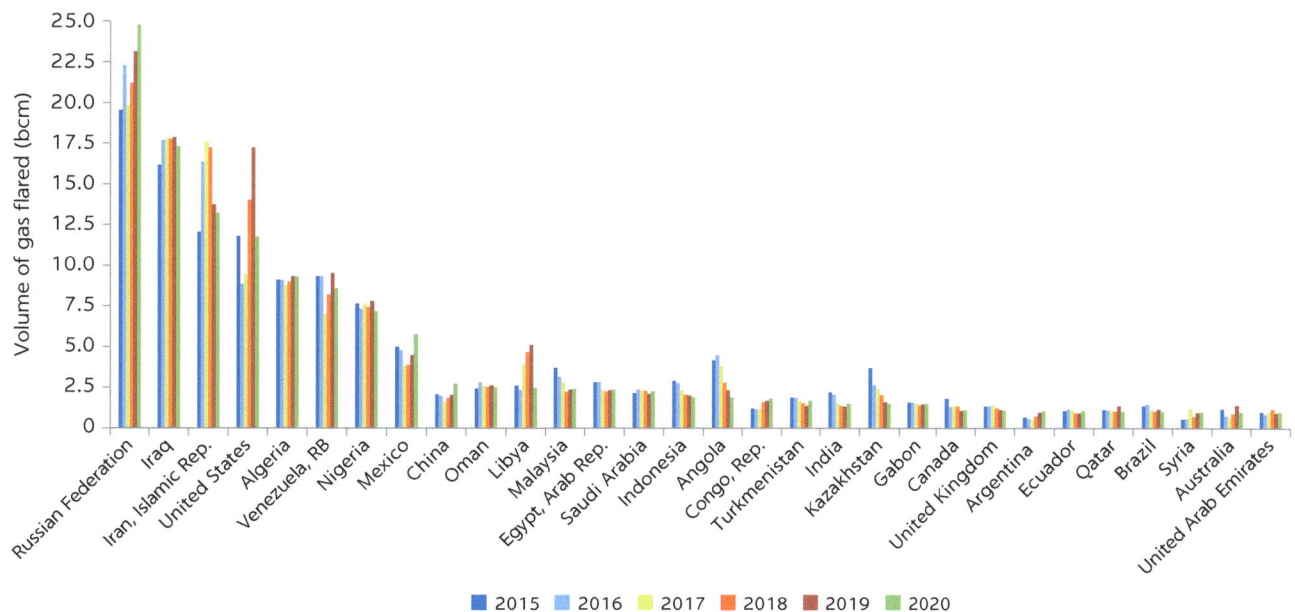

Source: Global Gas Flaring Reduction Partnership, based on data from US National Oceanic and Atmospheric Administration and Colorado School of Mines.
Note: Countries are ranked by flaring volume in 2020. bcm = billion cubic meters.

FIGURE 1.5

FIGURE 1.5

Global gas flaring and oil production, 1996–2020

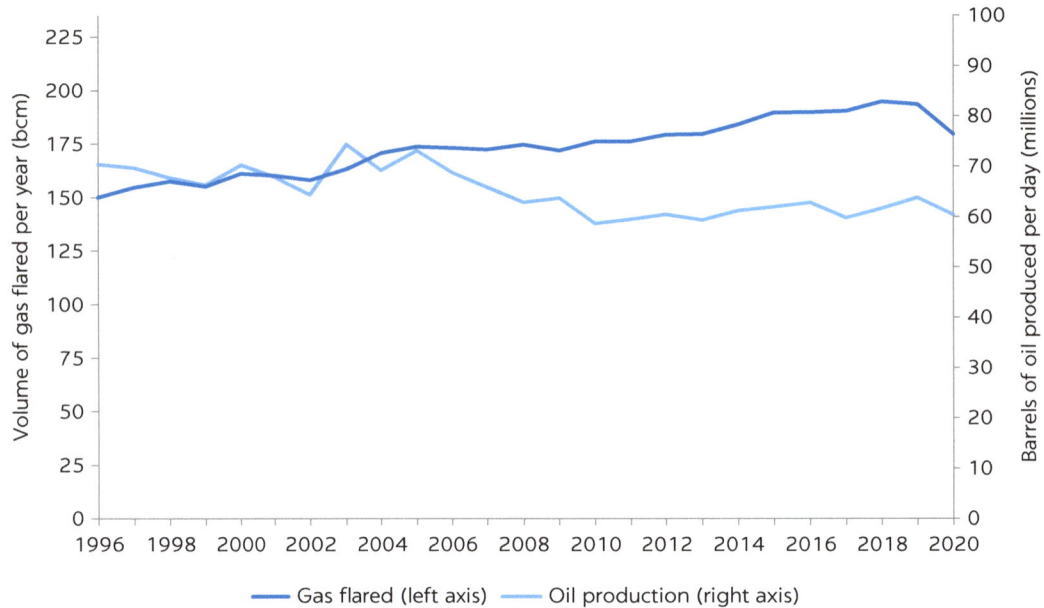

Source: Global Gas Flaring Reduction Partnership, based on data from the US National Oceanic and Atmospheric Administration, Colorado School of Mines, and US Energy Information Agency.
Note: bcm = billion cubic meters.

FIGURE 1.6

Flaring intensity of top 30 flaring countries, 2015–20

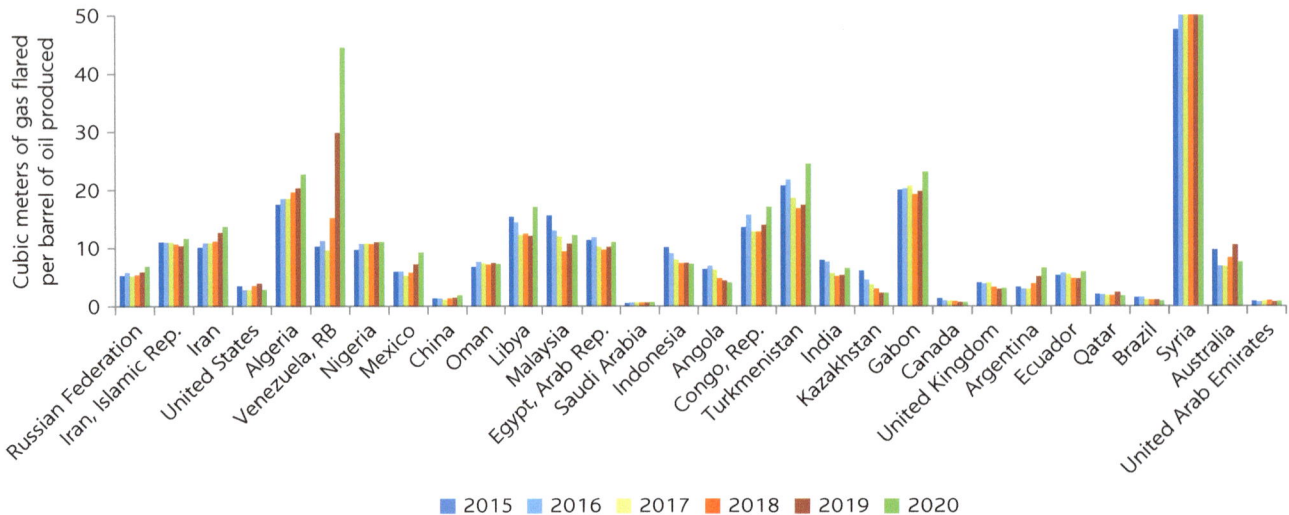

Source: Global Gas Flaring Reduction Partnership, based on data from the US National Oceanic and Atmospheric Administration, Colorado School of Mines, and US Energy Information Agency.
Note: Countries are ranked by 2020 flaring volume. Flaring intensity is measured in cubic meters of gas flared per barrel of oil produced.

Although the trend in flaring intensity can provide an indication of how well a country is managing its gas flaring over time, comparisons between the flaring intensity of different countries should be treated with great caution. The volume of associated gas produced per barrel of oil (the gas-to-oil ratio) can be very different between countries. Countries with a high gas-to-oil ratio have more gas to manage than those with low gas-to-oil ratios.

METHANE EMISSIONS

Key concepts

Methane emissions refer to the direct release of methane in the atmosphere without ignition, at different points along the oil and gas value chains. As previously noted, methane is a much more potent GHG than CO_2, with an estimated global warming potential of 28–36 over 100 years. Methane emissions fall under two categories (IEA 2020c):

1. *Vented emissions*, that is, emissions from the release of methane in the atmosphere occurring during normal operations due to the design of the facility or equipment, or for maintenance reasons.

2. *Fugitive emissions*, that is, accidental emissions due to, for example, equipment failure or poor maintenance. Fugitive emissions include those due to incomplete gas flaring because the combustion efficiency of gas flares, as previously noted, is never 100 percent.

Reasons and technical solutions

Both vented and fugitive emissions occur at different phases of the oil and gas value chains, with the four largest sources being onshore conventional oil and gas production, downstream gas processing, and unconventional gas production. Table 1.5 summarizes the different sources of and reasons for methane emissions in the oil and gas value chains. The following subsection contains a detailed analysis of the emission volumes presented in table 1.5.

TABLE 1.5 **Type and estimated volume of methane emissions from oil and gas, by phase, 2020**

INDUSTRY	TYPE OF EMISSIONS	BREAKDOWN BY PHASE	
		UPSTREAM	MID/DOWNSTREAM
Oil	Vented	22.5 MtVents for associated gas (in absence of flares); oil tank venting; leaking valves; leaking flanges; leaking oil pump seals; equipment venting during maintenance	Negligible
	Fugitive (excluding incomplete flares)	2 Mt Equipment failure; poorly maintained equipment and piping	Negligible
	Fugitive (incomplete flares)	3 Mt Oil field flare efficiency <100 %	n.a.
	Total	**27.5 Mt**	**Negligible**
Gas	Vented	18.2 Mt Gas field actuators; leaking valves; leaking flanges; leaking or inappropriate type of compressor seals; equipment venting during maintenance	5.5 MtGas pipeline actuators and controllers; leaking or inappropriate type of compressor seals; equipment venting during maintenance
	Fugitive (excluding incomplete flares)	10.8 Mt Equipment failure; poorly maintained equipment and piping	9.9 Mt Equipment failure; poorly maintained equipment and piping
	Fugitive (incomplete flares)	n.a.	n.a.
	Total[a]	**29 Mt**	**15.4 Mt**

Source: International Energy Agency Methane Tracker, https://www.iea.org/reports/methane-tracker-2021.
Note: Mt = million tons; n.a. = not applicable.
a. Total methane emissions from oil and gas were estimated at 72.1 Mt. Subtotals by category may not sum to 72.1 Mt because of rounding.

The Methane Guiding Principles, a voluntary partnership of oil and gas companies with 22 signatories at the time of writing, has identified best practices for methane emission reduction in eight areas, summarized in table 1.6. The partnership also published best-practice guides, which provide a summary of current known mitigations, costs, and available technologies to help those responsible for developing methane management plans.[14]

In general, methane emissions can be significantly reduced by implementing a series of technical and operational improvements to oil and gas activities. For instance, Beck et al. (2020) and the IEA (2020a) mention instrument air systems to replace pneumatic controllers to reduce venting, vapor recovery units installed on crude oil and condensate storage tanks, introducing leak detection and repair programs to significantly cut fugitive emissions, and applying the best available

TABLE 1.6 **Methane Guiding Principles: Best practices for methane emissions reduction**

PRIORITY AREA	BEST PRACTICES
Measurement and reporting of GHG emissions	• Develop a standardized industry-level data management system.
Engineering design and construction	• Use electric, mechanical, or instrument air-powered equipment where possible (including pneumatic controllers, pumps, and engines). • Have centralized and consolidated facilities where possible. • Use pipelines for liquid and gas takeaway. • Recover natural gas for beneficial use where possible. • Flare or combust natural gas when recovery is not possible. • Consider the use of alternative low-emission equipment/process. • Consider the use of alternative low-maintenance equipment/process.
Flaring	• Keep an accurate inventory of flaring activity. • Prevent flaring by designing systems that do not produce waste gases. • Recover waste gases as products to be sold. • Inject waste gases into oil or gas reservoirs. • Find alternative uses for flared gases, such as generating electricity. • Improve the efficiency of combustion when gases have to be flared. • Track progress in reducing flaring and venting.
Energy use	• Keep an accurate inventory of where natural gas is used as fuel. • Use electrical power or pneumatic power using compressed air or nitrogen. • Improve the energy efficiency of gathering operations and other equipment. • If natural gas needs to be used, improve the efficiency of fuel combustion. • Track progress in reducing fuel use.
Equipment leaks	• Keep an accurate inventory of emissions from equipment leaks. • Conduct a periodic leak detection and repair program. • Consider using alternative monitoring programs. • Replace or eliminate components that persistently leak.
Venting	• Keep an accurate inventory of venting activity. • Alter physical systems and operating practices to reduce venting. • Recapture gas where possible. • If methane needs to be released, flare it rather than venting it. • Track progress in reducing venting.
Pneumatic devices	• Keep an accurate inventory of pneumatic controllers and pumps powered by natural gas. • Replace pneumatic devices with electrical or mechanical devices where practical. • If pneumatic devices are used, eliminate emissions by using compressed air rather than natural gas to power them. • If using devices powered by natural gas is the best option, replace high-bleed controllers with alternatives with lower emissions. • Include pneumatic devices in an inspection and maintenance program, and report emissions from these devices in an annual inventory.

(continued next page)

TABLE 1.6, *continued*

PRIORITY AREA	BEST PRACTICES
Operational repairs	For operational repairs, • Perform periodic leak-detection surveys; • Repair leaks as soon as practical; • Check that repairs have been successful; • Keep track of repairs that have not been carried out; and • Keep and analyze records of leaks and repairs. For routine maintenance and repairs, • Use pumpdowns for pipelines and large vessels; • Minimize the volume that must be depressurized; • Use vapor-recovery units when pigging; • Avoid emissions by, for example, using hot taps to make connections to pipelines, carrying out nonintrusive inspections, and coordinating repairs and maintenance; and • If venting is necessary, flare the vented gases.
Continual improvement	• Commit to a program of methane management. • Improve methane reduction capabilities for preventing, identifying, and repairing leaks, and using effective engineering and design. • Set strong methane-reduction targets. • Report methane-reduction efforts and results. • Integrate methane management into the company culture.

Source: Based on Methane Guiding Principles, "Best Practice Guides" (https://methaneguidingprinciples.org/best-practice-guides/).

technologies (such as double mechanical seals on pumps, dry gas seals on compressors, and carbon packing ring sets on valve stems). The IEA (2020a) estimates that, if all these options were deployed across the oil and gas value chain, about 75 percent of the 2019 estimated 82 Mt of methane emissions from oil and gas operations could be avoided.

Global methane emission volumes

The concentration of methane in the atmosphere is about two-and-half times greater than preindustrial levels and is increasing. However, estimates of global methane emissions are subject to a great degree of uncertainty. The most comprehensive recent estimates are provided by the IEA (2020c) and put global methane emissions at 570 Mt, 40 percent of which originated from natural sources and 60 percent from human activity ("anthropogenic" emissions; figure 1.7).[15] The largest source of anthropogenic methane emissions is agriculture, responsible for about a quarter of the total, closely followed by the energy sector, which includes emissions from coal, oil, natural gas, and biofuels (IEA 2020c).

This uncertainty involves also, specifically, the estimates of oil and gas methane emissions, despite the emergence of new satellite data and other measurement tools. Several factors contribute to this uncertainty (IEA 2020b):

• Reporting companies mostly rely on average emission or activity factors, rather than on measured levels. The emission factors, derived from limited data, may not be applicable to all oil and gas facilities.

• Reported emissions may not be representative of those of the industry as a whole: companies that actively report methane emissions levels are generally the "best performers" in their peer group.

• Top-down studies misallocate emissions to the oil and gas sector, attributing to it methane emissions that in fact come from other sources such as coal, agriculture, or natural sources.

FIGURE 1.7
Sources of methane emissions

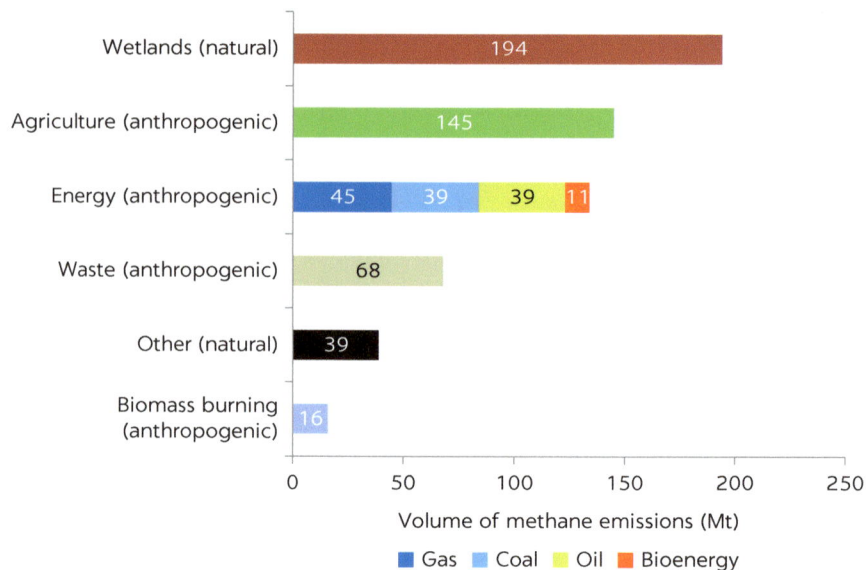

Source: IEA 2020c.
Note: Other natural sources include fresh water, geologic seepage, wild animals, termites, wildfires, permafrost, and vegetation. Mt = million tons.

With these caveats, the IEA estimates that 72.1 Mt of methane emissions derived from oil and gas operations in 2020, corresponding to 12.6 percent of global emissions and 21.1 percent of global human-made methane emissions.[16] Oil and gas operations are the largest source of methane emissions in the energy sector, which is collectively the second-largest source of anthropogenic methane emissions after agriculture. Energy-related methane emissions also come from coal and biofuels.[17] The IEA compared its estimates with those of several other institutions and academics and found them to be generally consistent (figure 1.8).[18]

Geographically, methane emissions are well distributed across oil and gas producing regions. One point of contention is the impact of the shale oil and gas boom in North America on methane emissions. Evidence analyzed by Howarth (2019), for instance, points to a correlation between the rise of methane in the atmosphere and the boom in fracking in the United States in the decade ending in 2019. Ambrose (2019), by contrast, notes that there is significant uncertainty around this theory.

According to the IEA, upstream operations are the largest source of methane emissions in the oil and gas sector, at about 56.4 Mt. Downstream methane emissions are about 15.4 Mt and are almost exclusively from gas operations (figure 1.9).[19]

The IEA makes the following estimates:[20]

- 64.2 percent of methane emissions from oil and gas operations (46.2 Mt) are the result of venting. Onshore conventional oil production is the main source of vented methane, followed by onshore conventional gas production. Unconventional and offshore production of oil and gas and downstream gas are also responsible for venting, but to a lesser extent.

FIGURE 1.8

Estimates of methane emissions from oil and gas

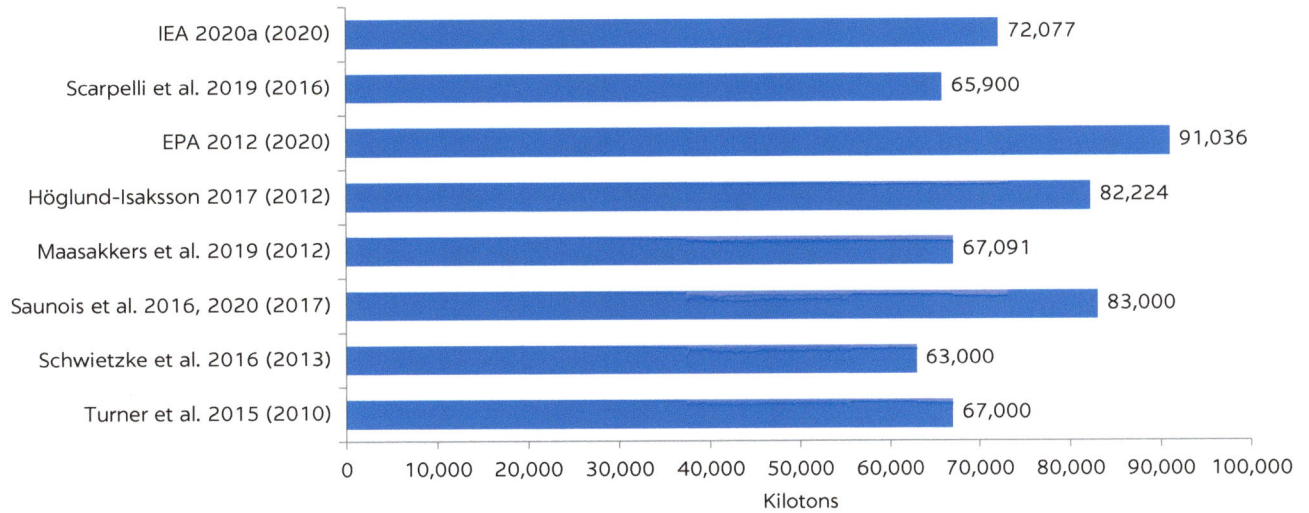

Source	Kilotons
IEA 2020a (2020)	72,077
Scarpelli et al. 2019 (2016)	65,900
EPA 2012 (2020)	91,036
Höglund-Isaksson 2017 (2012)	82,224
Maasakkers et al. 2019 (2012)	67,091
Saunois et al. 2016, 2020 (2017)	83,000
Schwietzke et al. 2016 (2013)	63,000
Turner et al. 2015 (2010)	67,000

Sources: International Energy Agency Methane Tracker Database (https://www.iea.org/articles/methane-tracker-database).
Note: Base year for each estimate in parentheses; see the sources at the end of this chapter.

FIGURE 1.9

Oil and gas methane emissions, by source, 2020

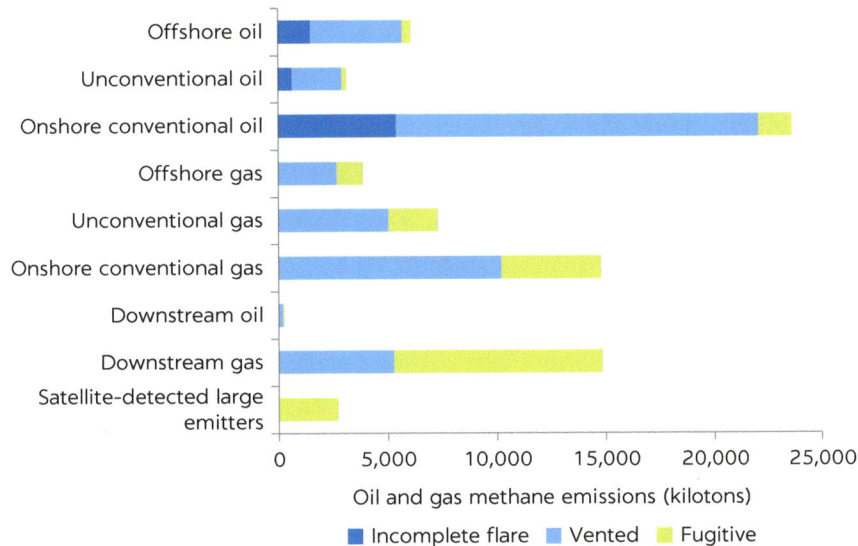

Legend: ■ Incomplete flare ■ Vented ■ Fugitive

Source: Based on International Energy Agency Methane Tracker Database (https://www.iea.org/articles/methane-tracker-database).
Note: Satellite-detected large emitters refer to fugitive emissions from upstream gas.

- 31.6 percent of methane emissions from oil and gas operations (22.8 Mt) are fugitive (excluding incomplete gas flaring). Downstream gas is the main source of fugitive emissions, followed by onshore conventional gas production, unconventional gas, and, to a lesser extent, onshore conventional oil production.

- 4.2 percent of methane emissions from oil and gas operations (over 3 Mt) occur because of incomplete gas flaring. As previously noted, flaring never achieves full combustion efficiency, and some volume of gas passes through unburned (typically about 2 percent in a well-designed flare, but often higher).

The IEA attributes the large portion of vented (as opposed to fugitive) emissions to (1) the broad definition of venting and (2) poor regulatory and enforcement standards in some countries, which result in operators deliberately venting rather than flaring, despite the former's much higher impact on global warming. Vented emissions, in the definition adopted by the IEA and this study, also include emissions resulting from equipment design; fugitive emissions are defined as, in essence, the result of unforeseen events. In addition, examining methane leakage rates for different types of oil and gas production facilities globally, the IEA has observed higher rates in countries with poor regulation, oversight, and enforcement—for example, in Central Asia. Some of the lowest leakage rates are in advanced oil economies such as Norway.

Similar to flaring, projections of future methane emissions are subject to a great degree of uncertainty. Long-term projections depend on a wide range of variables. In the IEA's Sustainable Development Scenario, global oil and gas methane emissions in 2040 would fall to less than 20 Mt (more than a 75 percent drop). This scenario, however, relies primarily on the assumption of a significant fall in oil and gas consumption by 2040, which would drive methane emissions down even without explicit abatement policies and measures. The IEA recognizes that relying solely on demand trends would be a "huge missed opportunity" in the efforts to mitigate climate change (IEA 2020a).[21]

REGULATORY DEVELOPMENTS

Review of gas flaring and venting regulation

A World Bank study of gas flaring regulation and regulatory practices in 28 oil-producing countries has produced the following key lessons of what constitutes an effective regulation (GGFR, forthcoming):

- The key approaches toward developing an effective gas flaring and venting regulation are prescriptive and performance-based:
  - The *prescriptive* approach is based on specific and detailed gas flaring and venting regulations, using detailed prescriptions of regulatory procedures and operational processes to make clear what is required.
  - The *performance-based* approach sets objectives and targets, leaving it to the operator to achieve these targets and demonstrate compliance.
- The effectiveness of the regulatory approach suitable for any given country depends largely on
  - The suitability of the methods and targets considered for the curtailment of routine flaring and venting, which result from an up-front consultation process with public and private sector stakeholders;
  - The availability and quality of measured results; and
  - Adequate enforcement capabilities.
- The relevance of flaring and venting across ministerial responsibilities can lead to unclear reporting lines, conflicting mandates, and reduced effectiveness of the regulatory agency. These issues apply particularly in countries with a dedicated ministry for oil and gas, but with flaring and venting falling under the responsibility of the ministry responsible for the environment. For example, in the US state of Colorado, a liaison officer has been put in

place for areas of joint interest between the various government agencies dealing with flaring and venting.

- The regulatory agency's monitoring and enforcement capabilities are key determinants for the eventual effectiveness of any regulation. A key element is to allocate monitoring and enforcement powers under one single agency, and to ensure that adequate measurement requirements are in place to provide the necessary data to enable targeted action from the regulator. To avoid these challenges, countries such as Norway have put in place a consultation process leading up to the development of their regulations, involving feedback loops from the agencies involved on the one hand and, where necessary, private sector participants on the other.

- Market-based solutions can be a valuable complementary measure to support the overall targets, because they use commercial considerations as an additional incentive. Several approaches have been implemented, some of which are still being assessed. Examples include

  - Kazakhstan's emissions trading system enabling oil and gas producers with emissions exceeding 20,000 $MtCO_2$/year to obtain the required emission quotas; and

  - Russia's preferential access of associated gas to the pipeline system, and preferential access of electricity produced from associated gas to the wholesale market.

Consequently, the regulatory approaches adopted, and their effectiveness, can vary widely, as illustrated by the examples in box 1.2.

Effective regulation can play a significant role in further curbing the wasteful practices of flaring and venting. To do so, it must create the right incentives to deploy the full spectrum of innovative technical or commercial solutions, and to collaborate across the industry.

Flaring and venting reduction and nationally determined contributions

The Paris Agreement is a legally binding international treaty on climate change, which entered into force at the end of 2016. Its goal is to limit global warming to well below 2.0 degrees Celsius, preferably to 1.5 degrees Celsius, compared to pre-industrial levels. To achieve this long-term temperature goal, countries aim to reach global peaking of GHG emissions as soon as possible, according to their capacity, to achieve a climate neutral world by midcentury.[22] Participating countries set their emissions reduction targets in the form of commitments known as Nationally Determined Contributions (NDCs). To date, only 13 countries have included gas flaring or methane reduction in their NDCs (table 1.7). Of those countries, 11 have endorsed the Zero Routing Flaring by 2030 initiative (see map 1.1 in the next subsection on international initiatives and voluntary standards).

An analysis of flaring volumes (figure 1.10) and flaring intensities (figure 1.11) for the countries listed in table 1.7 shows that progress in the last five years has been challenging across the board, although for different reasons. For example, República Bolivariana de Venezuela's sharp rise in flare intensity evidences the country's difficulties in dealing with ailing infrastructure, and Nigeria, which steadily reduced its flaring volumes by some 70 percent over the past 15 years, is now dealing with the challenge of bringing small flares to the market (GGFR 2021).[23]

BOX 1.2

Examples of countries' regulatory approaches to gas flaring

Norway is the world's fifth-largest crude exporter, and among the major oil-producing countries. It also has the lowest flaring rates. According to data from the Global Gas Flaring Reduction Partnership of the World Bank, Norway's flare gas volumes decreased from 0.3 billion cubic meters (bcm) in 2012 to 0.1 bcm in 2020. The last flare count in 2019 found just 28 individual flare sites.

Norway has imposed restrictions on flaring and venting since oil production began in the early 1970s, and the country's current regulatory framework relies on the following main pillars:

- All unauthorized or not justified (from a safety perspective) flaring and venting are subject to a fine.
- Emissions from flaring and venting are subject to a carbon tax.
- The country is also part of the European Union's Emissions Trading System.

The Russian Federation is the largest contributor of flare gas by volume. According to data from the Global Gas Flaring Reduction Partnership, Russian flare volumes increased from 22.4 bcm in 2012 to 24.9 bcm in 2020. Furthermore, the last flare count in 2019 found 1,086 individual flare sites.

Starting with the Energy Strategy of Russia in 2009, the country started to issue laws and regulations limiting flaring and venting:

- It placed a general limitation on routine and nonroutine flaring to 5 percent of the associated gas produced in combination with volume-based fee payments.
- The fees increase substantially in case of deviation (for example, surcharges for surpassing the 5 percent threshold, or if no metering system is in place).
- It incentivizes the use of associated gas by providing preferential access to the national pipeline system and the power market (for electricity produced from associated gas).

Malaysia ranks 12th globally in terms of gas flared. According to data from the Global Gas Flaring Reduction Partnership, flare gas volumes increased from 2.3 bcm in 2012 to a peak of about 3.7 bcm in 2015, before falling to 2.4 bcm in 2020.

In contrast to most other countries, Malaysia delegated most of its flaring and venting prevention and enforcement responsibilities to its state-owned oil company Petronas:

- There is no detailed formal regulation covering flaring and venting, but Petronas is tasked to ensure compliance with their (nonpublic) operating procedures, including the resourceful use of associated gas.
- Noncompliance can be subject to fines, but there is no publicly available framework stipulating the fee amounts and procedure.

Source: GGFR, forthcoming.

The impact of a reduction of associated gas flaring on NDC targets varies greatly by country. Elvidge et al. (2018) find that, globally, flaring reductions could provide less than 2 percent of the emission reductions presented in the NDCs. At the country level the potential to meet NDC targets varies as a function of the flared gas volume, total projected emissions, and the NDC goal as a percentage of total projected emissions. The authors suggest that (1) countries that can meet more than 10 percent of their NDC targets from gas flaring should have a high incentive to invest in it, (2) countries that can exceed their NDC target through gas flaring reductions may consider increasing their NDC reduction targets, and (3) countries that have yet to establish NDC targets can use similar estimates as input to their NDC deliberations.

TABLE 1.7 **Countries that included flaring reduction in their Nationally Determined Contributions**

COUNTRY	YEAR	FLARING REDUCTION COMMITMENT
Algeria	2016	Reduce the volume of gas flaring to less than 1 percent by 2030.[a]
Angola	2021	Reduce flaring—295 mmscf/d (42 percent of unconditional commitments) or 370 mmscf/d considering conditional commitments compared to 2015 levels.
Bahrain	2021	Undertake projects that reduce CO_2 emissions, including flaring and associated gas compression projects. However, no specific emissions reduction target is provided.
China	2021	Enhance the recovery and use of vented and associated natural gas. However, no specific measure or target is provided.
Congo, Rep.	2021	Flaring is estimated to account for 23 percent of direct GHG emissions from the energy sector in 2000. The first NDC refers to various measures of flaring being taken over the years, and a policy to encourage its productive use when reinjection is not possible. However, no specific emissions reduction target is provided.
Ecuador	2019	Reduce gas flaring and use associated gas in power generation. However, no specific emissions reduction target is provided.
Egypt, Arab Rep.	2017	Undertake GHG emissions reduction in the oil and natural gas sector, including venting and flaring through the use of advanced locally appropriate and more-efficient fossil fuel technologies, which emit less. However, no specific emissions reduction target is provided.
Gabon	2016	Over 2010–25, policies will reduce GHG emissions from flaring by an estimated 17,341 $GtCO_2e$, or 41 percent of emissions (63 percent in 2025). Actions toward this goal include investments in reinjection, gas flare to power, and compression units.
Iraq	2021	Reduce the levels of burning associated gas and invest it in oil and natural gas extraction processes. Conduct regular detection programs for methane leaks at oil and gas facilities for repair. However, no specific emissions reduction target is provided.
Mexico	2020	Actions that will promote the optimization of the processes of the refining and processing systems have been identified for the oil and gas sector, including the implementation of the Methane Emissions Reduction Policy. However, no specific sectoral target has been proposed.
Oman	2019	Oman would reduce its expected GHG growth through several mitigation contributions, including the reduction of gas flaring from oil industries. However, no specific emissions reduction target is provided.
Qatar	2021	Qatar committed to Zero Routine Flaring by 2030, with a long-term goal to reduce flaring in onshore facilities to the absolute minimum. Technically feasible nonroutine flaring is also covered. CCS will be another pivotal measure for reducing the sector's GHG emissions. Qatar commissioned the largest CO_2 recovery and sequestration facility in the MENA region at Ras Laffan in 2019 with a design capacity of 2.2 MTPA of CO_2 capture and storage. CCS is included for new LNG facilities and concepts, for remaining LNG facilities will be developed and implemented considering the economic, safety and environmental concerns. However, no specific emissions reduction target is provided.
Saudi Arabia	2021	Zero flaring by 2030. Reduce global methane emissions by 30 percent by 2030 relative to 2020 levels.
United Arab Emirates	2020	Having adopted Zero Routine Flaring policies, the country's oil and gas companies are adopting measures to avoid flaring, use recovered gas in operations, and develop CCUS. In its second NDC submitted in 2020, the country noted the Abu Dhabi National Oil Company's target to decrease its GHG emission intensity by 25 percent by 2030. In 2016 the region's first commercial-scale network for CCUS was developed in Al Reyadah. Advanced techniques are also deployed to measure and control methane leaks. No specific sector level target is proposed.
Venezuela, RB	2015	The country's oil and gas industry is working toward minimizing gas flaring and venting. These measures were estimated to reduce emissions by 538.2 $ktCO_2e$/year for the period 2016–19.

Source: NDC Registry, United Nations Framework Convention on Climate Change (https://www4.unfccc.int/sites/NDCStaging/Pages/All.aspx), last accessed November 9, 2021.
Note: CCS = carbon capture storage; CCUS = carbon capture, utilization, and storage; CO_2 = carbon dioxide; GHG = greenhouse gas; $GtCO_2e$ = gigatons of CO_2 equivalent; $ktCO_2e$ = kilotons of CO_2 equivalent; LNG = liquefied natural gas; MENA = Middle East and North Africa; mmscf/d = million standard cubic feet per day; MTPA = million tons per annum; NDC = Nationally Determined Contribution.
a. Baseline year missing from Algeria's NDC submission.

FIGURE 1.10

Flaring volumes in countries with flare reduction in their Nationally Determined Contributions, 2015–20

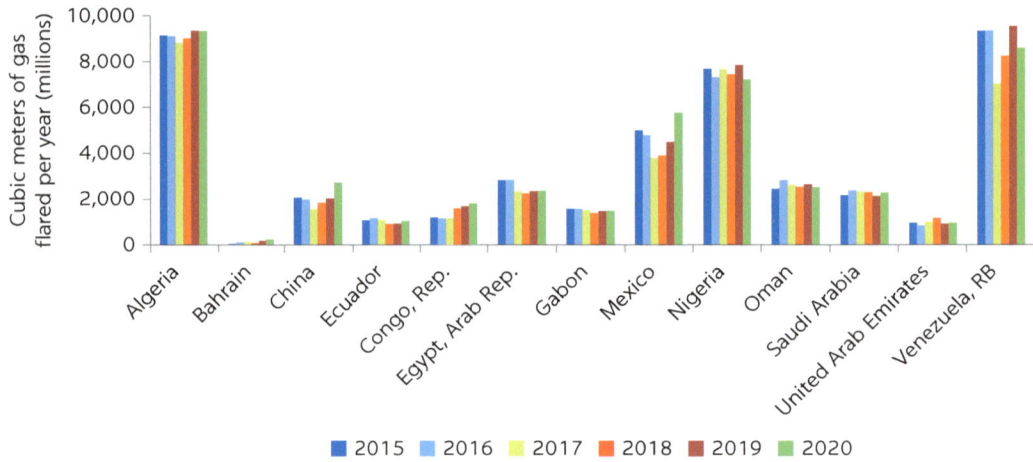

Source: Based on data provided by the Global Gas Flaring Reduction Partnership.

FIGURE 1.11

Flaring intensity in countries with flare reduction in their Nationally Determined Contributions, 2015–20

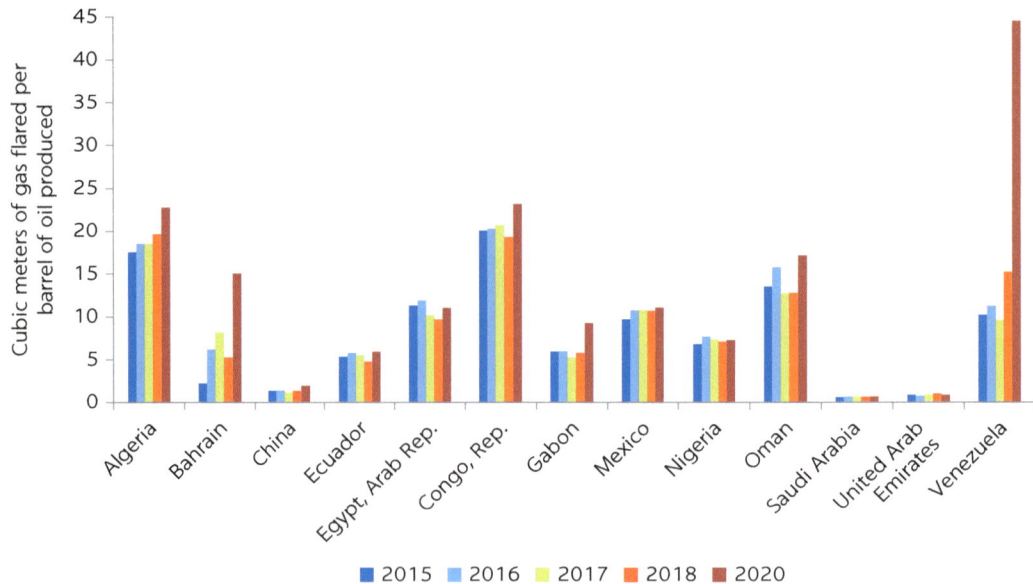

Source: Based on data provided by the Global Gas Flaring Reduction Partnership.

International initiatives and voluntary standards

A growing number of international initiatives focus on reducing gas flaring and methane emissions, highlighting the increasing importance attributed to this effort by governments, oil and gas companies, and international institutions. The main initiatives are briefly described here:

• Under the leadership of the World Bank Group, GGFR—a public-private initiative comprising international and national oil companies, national and regional governments, and international institutions—works to increase the use of associated gas by helping remove technical and regulatory barriers to flaring reduction, conducting research, disseminating best practices, and developing country-specific gas flaring reduction programs.

- The World Bank introduced the Zero Routine Flaring by 2030 initiative in 2015, with the objective to bring together governments, oil companies, and development institutions that agree to cooperate to eliminate routine flaring no later than 2030. Oil companies that endorse the initiative will develop new oil fields according to plans that incorporate sustainable use or conservation of associated gas without routine flaring. Oil companies with routine flaring at existing oil fields will seek to implement economically viable solutions to eliminate this legacy flaring as soon as possible, and no later than 2030. As of May 2020, 32 governments, 41 oil companies, and 15 development institutions have endorsed the initiative (map 1.1).[24]

- The Global Methane Initiative, launched in 2004, is an international public-private initiative that advances cost-effective, near-term methane abatement and recovery measures and the use of methane as a clean energy source in three sectors: biogas, coal mines, and oil and gas. The initiative comprises 45 governments and a wide range of project networks dedicated to tackling specific GHG reduction challenges.[25]

MAP 1.1

Zero Routine Flaring by 2030 endorsers

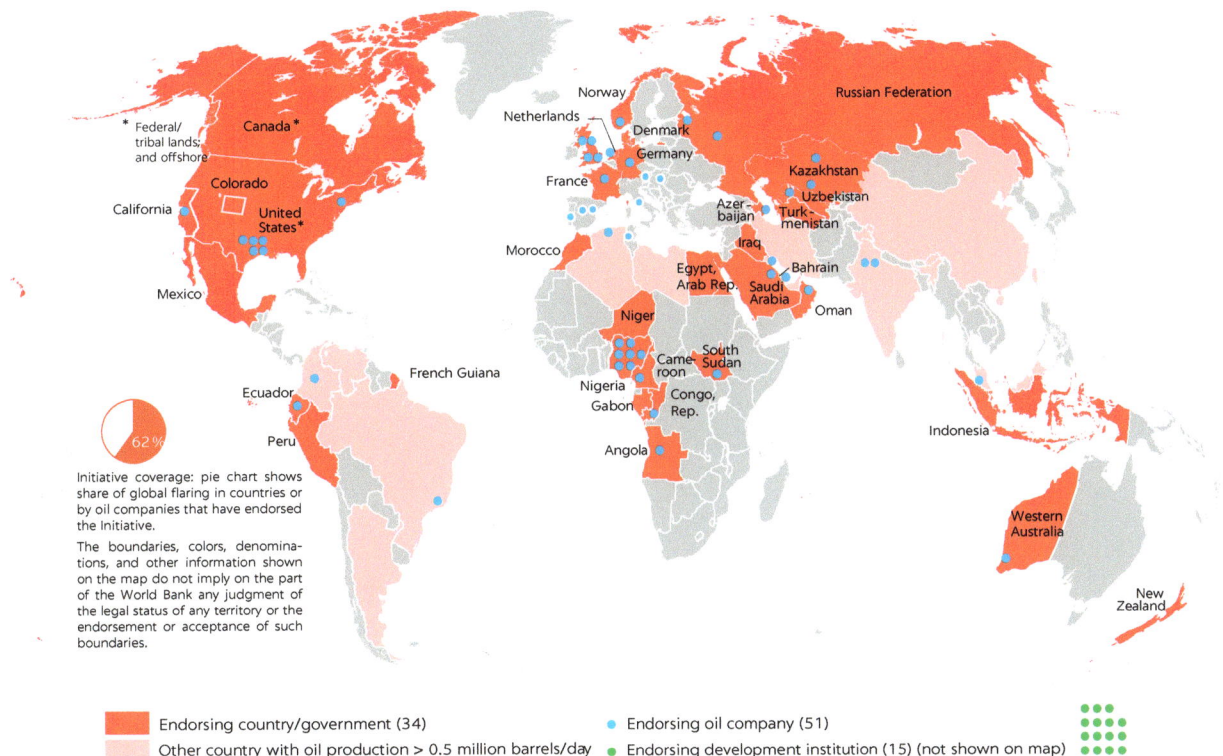

Initiative coverage: pie chart shows share of global flaring in countries or by oil companies that have endorsed the Initiative.

The boundaries, colors, denominations, and other information shown on the map do not imply on the part of the World Bank any judgment of the legal status of any territory or the endorsement or acceptance of such boundaries.

▮ Endorsing country/government (34)	● Endorsing oil company (51)
▮ Other country with oil production > 0.5 million barrels/day	● Endorsing development institution (15) (not shown on map)

Source: Global Gas Flaring Reduction Partnership's Zero Routine Flaring website, accessed May 5, 2021 (https://www.worldbank.org/en/programs/zero-routine-flaring-by-2030#4).

- The Oil and Gas Methane Partnership, launched at the United Nations Climate Summit in 2014, was created by the Climate and Clean Air Coalition as a voluntary initiative to help companies reduce methane emissions in the oil and gas sector. Led by the United Nations Environment Programme, the European Commission, and the Environmental Defense Fund, the partnership is a multistakeholder group comprising 62 companies that are responsible for 30 percent of the world's oil and gas production. The partnership focuses on methane emissions reporting. It provides a protocol to help companies systematically manage their methane emissions from oil and gas operations, and offers a credible platform to help member companies demonstrate actual reductions to industry stakeholders. To support the realization of global climate targets, the Oil and Gas Methane Partnership 2.0 aims to deliver a 45 percent reduction in the industry's methane emissions by 2025, and a 60–75 percent reduction by 2030.[26]

- At the 2019 United Nations Climate Summit, the Climate and Clean Air Coalition called on governments to join a Global Methane Alliance to significantly reduce methane emissions from the oil and gas sector, as a way for countries to accelerate progress toward their NDCs. Countries that join the alliance will commit to either absolute methane reduction targets of at least 45 percent by 2025 and 60–75 percent by 2030 or to a near zero methane intensity target. As of May 2021, the alliance comprises 181 members, including states, intergovernmental organizations, and nongovernmental organizations.[27]

- In 2017, eight international oil and gas companies committed on a voluntary basis to five Methane Guiding Principles for the reduction of methane emissions along the natural gas value chain. The number of signatories subsequently grew to 24 at the time of writing (including most oil and gas majors). The five principles are (1) continually reduce methane emissions, (2) advance strong performance across the gas supply chain, (3) improve accuracy of methane emissions data, (4) advocate sound policy and regulations on methane emissions, and (5) increase transparency. As previously noted, the principles were also translated into eight best-practice guidelines for methane emission reduction.

- On October 14, 2020, the European Commission adopted the European Union Methane Strategy as part of the European Green Deal.[28] The strategy sets out measures to cut methane emissions in Europe and internationally. It presents legislative and nonlegislative actions in the energy, agriculture, and waste sectors, which combined account for about 95 percent of methane emissions associated with human activity worldwide.

- Ipieca, a nonprofit oil and gas industry association, included emission mitigation strategies and activities as suggested topics for disclosure by oil and gas operators in its Climate Change Reporting Framework. Disclosure would include the company's approaches to flaring reduction, venting and fugitive emissions reduction, and an explanation on whether or how methane is being considered in the emissions management plan (Ipieca 2019).

In addition to these international initiatives, several oil and gas groups have announced targets or plans to reduce scope 1 and scope 2 emissions. Targets are expressed in terms of total emission reductions or reductions in emissions intensity. Plans vary from commitments expressly incorporated in business plans to more aspirational ones (IEA 2020b). Table 1.8 summarizes the initiatives of several prominent oil and gas operators.

TABLE 1.8 Emission reduction commitments and targets of selected companies

COMPANY	METHANE EMISSION TARGET	GHG EMISSION TARGET
BP[a]	Reduce methane intensity of operations by 50% by 2023	Reduce scope 1 and scope 2 emissions by 3.4 MtCO$_2$e for the period 2015–25; net zero on its oil and gas operations by 2050; 50% cut in the carbon intensity of products BP sells by 2050 or sooner; install methane measurement at all existing major oil and gas processing sites by 2023, publish the data, and then drive a 50% reduction in methane intensity of its operations.
Chevron[b]	Reduce methane emissions intensity by 53% (2016–28)	Reduce upstream oil and gas net GHG emission intensity by 40% and 26%, respectively (2016–28); net zero (scope 1 and scope 2) by 2050.
Eni[c]	Target of 80% reduction of fugitive methane emissions from 2014 level by 2025	Reduce net GHG life-cycle emission (scopes 1, 2, and 3) intensity by 25% in 2030 vs. 2018; net-zero GHG life-cycle emissions by 2050.
ExxonMobil[d]	Reduce methane intensity by 40–50% from 2016 levels by 2025	15–20% reduction in GHG intensity in upstream operations by 2025 (based on 2016 levels) covering scope 1 and scope 2.
Shell[e]	Target to maintain < 0.2% intensity by 2025	Zero routing flaring of gas by 2025; 20% reduction in carbon intensity by 2030 (including scope 1, 2, and 3 emissions) based on 2016 levels; 50% reduction in absolute emissions by 2030 (including scope 1, 2, and 3 emissions) on a net carbon footprint basis compared to 2016 levels.
TotalEnergies[f]	Target to maintain < 0.2% intensity of operated oil and gas assets and < 0.1% intensity of operated gas assets by 2030	Zero routing flaring by 2030; cut scope 1 and scope 2 emissions by 15% by 2025 and by 40% by 2030 compared to 2015 levels; carbon neutrality by 2050.
CNPC[g]	50% reduction in methane intensity by 2025 (from a 2019 baseline)	Zero Routine Flaring by 2030; "near zero" emissions around 2050.
Equinor[h]	Target intensity from 0.03% to near zero in Norway by 2030	Reduce emissions from its domestic operations by 40% by 2030, and to net zero by 2050 (including scope 1, 2, and 3 emissions).
Petrobras[i]	Reduce methane emissions intensity by 30–50% in upstream operations from 2015 levels by 2025	Reduce emissions by 32% in upstream operations and by 16% in refining from 2015 levels by 2025.
Petrona[j]	Zero continuous venting in upstream facilities in Malaysia by 2024	Zero routing flaring by 2030; net zero emissions by 2050.
Repsol[k]	Target < 0.2% intensity by 2024	Zero Routine Flaring by 2030; reduce emissions from operated assets (scope 1 and 2) by 55% and net emissions (scope 1, 2, and 3) by 30% by 2030 compared to 2016 levels; achieve net zero emissions by 2050.

Sources: World Bank, based on company websites:
a. BP (https://www.bp.com/en/global/corporate/sustainability/climate-change/reducing-emissions-in-our-operations.html and https://www.bp.com/en/global/corporate/news-and-insights/press-releases/bernard-looney-announces-new-ambition-for-bp.html).
b. Chevron (https://www.chevron.com/sustainability/environment/lowering-carbon-intensity).
c. Eni (https://www.eni.com/en-IT/low-carbon/ghg-emission-reduction.html).
d. ExxonMobil (https://corporate.exxonmobil.com/Sustainability/Emissions-and-climate).
e. Shell (https://www.shell.com/energy-and-innovation/natural-gas/methane-emissions.htm and https://www.shell.com/energy-and-innovation/the-energy-future/what-is-shells-net-carbon-footprint-ambition/faq.html).
f. TotalEnergies (https://totalenergies.com/sites/g/files/nytnzq121/files/documents/2021-10/TotalEnergies_Climate_Targets_2030_EN.pdf).
g. CNPC (https://www.cnpc.com.cn/en/climate/common_index.shtml and https://www.cnpc.com.cn/en/environmentcase/202111/3c9060a87d704600b63f0232a6f16962.shtml).
h. Equinor (https://www.equinor.com/en/news/20201102-emissions.html).
i. Petrobras (https://petrobras.com.br/en/society-and-environment/environment/climate-changes/).
j. Petronas (https://methaneguidingprinciples.org/wp-content/uploads/2021/01/Methane-Guiding-Principles_Reporting-Petronas.pdf and https://www.petronas.com/sustainability/net-zero-carbon-emissions).
k. Repsol (https://www.repsol.com/en/sustainability/climate-change/net-zero-emissions-2050/index.cshtml).

NOTES

1. Global human-made emissions were estimated at 49.04 $GtCO_2e$ in 2015, according the European Commission's Emissions Database for Global Atmospheric Research (EDGAR), accessed May 21, 2021 (https://edgar.jrc.ec.europa.eu/country_profile).
2. Based on 2015 global GHG emissions reported in EDGAR.
3. Regulation has proven effective in tackling scope 3 emissions, for instance by tightening vehicle emission standards.
4. Based on 159 billion cubic meters of natural gas flared in 2019, 98 percent flare efficiency, and a global warming potential of methane of 30 on a 100-year horizon basis. Throughout this report, methane is assigned a global warming potential of 30 to ensure consistency with the IEA's (2021b) key assumptions.
5. Methane emissions during flaring are included in the previously mentioned 462 $MtCO_2e$ emissions from flaring.
6. Estimates of flared volumes for 2020 and 2019 come from the Global Gas Flaring Reduction Partnership (GGFR) website (https://www.worldbank.org/en/programs/gasflaringreduction#7). Estimates of methane emission volumes come from the IEA (2020b) and the IEA's Methane Tracker Database (https://www.iea.org/articles/methane-tracker-database).
7. In gas fields, there is no or minimal flaring. Relatively few gas fields contain gas with significant volumes of heavier hydrocarbons that are extracted from the produced gas and sold. In many cases the lighter components, methane and ethane, are reinjected into the reservoir. In some cases, however, these components are flared. See Elvidge et al. (2018) and the World Bank's Zero Routine Flaring by 2030 web page (https://www.worldbank.org/en/programs/zero-routine-flaring-by-2030#7).
8. The volume of gas flared in 2020 was equivalent to approximately 3.9 percent of the total gas produced globally.
9. World Bank's Zero Routine Flaring by 2030 web page (https://www.worldbank.org/en/programs/zero-routine-flaring-by-2030).
10. IEA's Methane Tracker Database (https://www.iea.org/articles/methane-tracker-database). Note that estimates of volume of gas vented are highly dependent on the gas density (kilogram per cubic meter) assumed.
11. The US Energy Information Agency, using 2012 data, estimates that 15 percent of associated gas production was flared or vented, 58 percent reinjected, and 27 percent used.
12. The calculation assumes a 40 percent conversion efficiency from gas to power. For medium- and large-scale power generation units, the conversion efficiency would be between 40 percent and 45 percent; small-scale units would likely show efficiencies between 30 percent and 35 percent.
13. The ideal measure of effectiveness of a country's gas use is the ratio of gas used to gas produced; however, volumes of associated gas produced are rarely publicly available.
14. The Methane Guiding Principles best-practice guides can be accessed at https://methaneguidingprinciples.org/best-practice-guides/.
15. This estimate is consistent with Saunois et al. (2016, 2020).
16. IEA's Methane Tracker Database (https://www.iea.org/articles/methane-tracker-database).
17. The IEA compared its estimates to several other sources and found them to be generally in line (https://www.iea.org/articles/methane-tracker-database).
18. IEA's Methane Tracker Database (https://www.iea.org/articles/methane-tracker-database).
19. IEA's Methane Tracker Database (https://www.iea.org/articles/methane-tracker-database).
20. IEA's Methane Tracker Database (https://www.iea.org/articles/methane-tracker-database).
21. Per the IEA (2021b), methane emissions from fossil fuel production and use fall from 115 Mt in 2020 (3.5 $GtCO_2eq$) to 30 Mt in 2030 and 10 Mt in 2050 in the Net Zero Emissions scenario. About one-third of this decline results from an overall reduction in fossil fuel consumption, but the larger share comes from a huge increase in the deployment of emission reduction measures and technologies, which lead to the elimination of all technically avoidable methane emissions by 2030 (IEA 2020a).
22. Of the 197 countries that endorsed the Paris Agreement, 190 have formalized their commitment. A handful of high emitters have yet to formally join the Paris Agreement, including the Islamic Republic of Iran, Iraq, and Turkey. For details on the Paris Agreement, see the United Nations Climate Change web page (https://unfccc.int/process-and-meetings/the-paris-agreement/the-paris-agreement).

23. For a detailed analysis of flaring and flaring intensity trends, see GGFR (2021).
24. GGFR's Zero Routine Flaring website, accessed May 5, 2021 (https://www.worldbank.org/en/programs/zero-routine-flaring-by-2030#4).
25. Information on the Global Methane Initiatives can be found at https://www.globalmethane.org/about/index.aspx, last accessed on May 25, 2021.
26. Additional information on the Oil and Gas Methane Partnership can be found at https://www.unep.org/news-and-stories/press-release/oil-and-gas-industry-commits-new-framework-monitor-report-and-reduce, last accessed May 25, 2021.
27. Climate and Clean Air Coalition website, accessed May 25, 2021 (https://www.ccacoalition.org/en/partners).
28. The strategy and the European Union Methane Target Plan are available at https://ec.europa.eu/commission/presscorner/detail/en/IP_20_1833, accessed May 25, 2021.

REFERENCES

Ambrose, Jillian. 2019. "Fracking Causing Rise in Methane Emissions, Study Finds." *Guardian*, August 14. https://www.theguardian.com/environment/2019/aug/14/fracking-causing-rise-in-methane-emissions-study-finds.

Argonne Venting and Flaring Research Team. 2017. "Analysis of Potential Opportunities to Reduce Venting and Flaring on the OCS." Bureau of Safety and Environmental Enforcement, US Department of Energy. https://www.bsee.gov/sites/bsee.gov/files/5007aa.pdf.

Beck, Chantal, Sahar Rashidbeigi, Occo Roelofsen, and Eveline Speelman. 2020. "The Future Is Now: How Oil and Gas Companies Can Decarbonize." McKinsey & Co., January 7. https://www.mckinsey.com/industries/oil-and-gas/our-insights/the-future-is-now-how-oil-and-gas-companies-can-decarbonize.

Buzcu-Guven, Birnur, and Robert Harriss. 2012. "Extent, Impacts and Remedies of Global Gas Flaring and Venting." *Carbon Management* 3 (1): 95–108.

Elvidge, Christopher D., Morgan D. Bazilian, Mikhail Zhizhin, Tilottama Ghosh, Kimberly Baugh, and Feng-Chi Hsu. 2018. "The Potential Role of Natural Gas Flaring in Meeting Greenhouse Gas Mitigation Targets." *Energy Strategy Reviews* 20 (2018): 156–62.

EPA (US Environmental Protection Agency). 2012. "Global Anthropogenic Non-CO_2 Greenhouse Gas Emissions: 1990–2030." Office of Atmospheric Programs, Climate Change Division, US EPA, Washington, DC.

GGFR (Global Gas Flaring Reduction Partnership). 2016. "Gas Flaring Definitions." World Bank, Washington, DC. http://documents.worldbank.org/curated/en/755071467695306362/pdf/106662-NEWS-PUBLIC-GFR-Gas-Flaring-Definitions-29-June-2016.pdf.

GGFR (Global Gas Flaring Reduction Partnership). 2019a. "Gas Flaring Estimates. Methodology for Determining the Flare Volumes from Satellite Data." World Bank, Washington, DC. http://pubdocs.worldbank.org/en/853661587048977000/Estimation-of-flare-gas-volumes-from-satellite-data-002.pdf.

GGFR (Global Gas Flaring Reduction Partnership). 2019b. "GGFR Technology Overview—Utilization of Small-Scale Associated Gas." World Bank, Washington, DC. http://documents.worldbank.org/curated/en/469561534950044964/pdf/GGFR-Technology-Overview-Utilization-of-Small-Scale-Associated-Gas.pdf.

GGFR (Global Gas Flaring Reduction Partnership). 2021. "Global Gas Flaring Tracker Report." World Bank, Washington, DC. https://thedocs.worldbank.org/en/doc/1f7221545bf1b7c89b850dd85cb409b0-0400072021/original/WB-GGFR-Report-Design-05a.pdf.

GGFR (Global Gas Flaring Reduction Partnership). Forthcoming. "Global Review of Regulation of Gas Flaring and Venting." World Bank, Washington, DC.

Hajilary, Nasibeh, Mahallah Rezakazemi, and Aref Shahi. 2020. "CO_2 Emission Reduction by Zero Flaring Startup in Gas Refinery." *Materials Science for Energy Technologies* 3 (2020): 218–24. https://reader.elsevier.com/reader/sd/pii/S2589299119301375?token=BF435D6FB7891E604E6C83390A3B0C4CFC74E1903CBC31F10AE032C8ACCA43B14BE867EE2590F270162A9F54055C5C8B&originRegion=us-east-1&originCreation=20210513182848.

Höglund-Isaksson, L. 2017. "Bottom-Up Simulations of Methane and Ethane Emissions from Global Oil and Gas Systems 1980 to 2012." *Environmental Research Letters* 12 (2).

Howarth, Robert W. 2019. "Ideas and Perspectives: Is Shale Gas a Major Driver of Recent Increase in Global Atmospheric Methane?" *Biogeosciences* 16 (15): 3033–46. https://doi .org/10.5194/bg-16-3033-2019.

IEA (International Energy Agency). 2018. *World Energy Outlook 2018*. Paris: IEA. https://iea .blob.core.windows.net/assets/77ecf96c-5f4b-4d0d-9d93-d81b938217cb/World_Energy _Outlook_2018.pdf.

IEA (International Energy Agency). 2020a. "The Oil and Gas Industry in Energy Transitions." World Energy Outlook Special Report. IEA, Paris. https://www.iea.org/reports/the -oil-and-gas-industry-in-energy-transitions.

IEA (International Energy Agency). 2020b. "Global Methane Emissions from Oil and Gas: Insights from the Updated IEA Methane Tracker." IEA, Paris. https://www.iea.org/articles /global-methane-emissions-from-oil-and-gas.

IEA (International Energy Agency). 2020c. "Methane Tracker 2020." IEA, Paris. https://www .iea.org/reports/methane-tracker-2020.

IEA (International Energy Agency). 2020d. "Tracking Fuel Supply 2020." IEA, Paris. https:// www.iea.org/reports/tracking-fuel-supply-2020.

IEA (International Energy Agency). 2020e. "The Covid-19 Crisis and Clean Energy Progress." IEA, Paris. https://www.iea.org/reports/the-covid-19-crisis-and-clean-energy-progress.

IEA (International Energy Agency). 2021a. "Global Energy Review: CO_2 Emissions in 2020." IEA, March 2, 2021. https://www.iea.org/articles/global-energy-review-co2-emissions-in-2020.

IEA (International Energy Agency). 2021b. "Net Zero by 2050. A Roadmap for the Energy Sector." IEA, Paris. https://iea.blob.core.windows.net/assets/ad0d4830-bd7e-47b6-838c -40d115733c13/NetZeroby2050-ARoadmapfortheGlobalEnergySector.pdf.

Ipieca. 2018. "Methane Glossary." Ipieca, London. https://www.ipieca.org/resources/awareness -briefing/methane-glossary/.

Ipieca. 2019. "Ipieca Climate Change Reporting Framework: Supplementary Guidance for the Oil and Gas Industry on Voluntary Sustainability Reporting." Updated May, Ipieca, London. https://www.ipieca.org/resources/good-practice/ipieca-climate-change -reporting-framework-supplementary-guidance/.

Maasakkers, J., D. Jacob, M. Sulprizio, T. Scarpelli, H. Nesser, J.-X. Sheng, Y. Zheng, et al. 2019. "Global Distribution of Methane Emissions, Emission Trends, and OH Concentrations and Trends Inferred from an Inversion of GOSAT Satellite Data for 2010–2015." *Atmospheric Chemistry and Physics* 19 (11): 7859–81.

Olivier, J. G. J., and J. A. H. W. Peters. 2018. "Trends in Global CO_2 and Total Greenhouse Gas Emissions: 2018 Report." PBL Netherlands Environmental Assessment Agency, The Hague.

Saunois, Marielle, Philippe Bousquet, Ben Poulter, Anna Peregon, Philippe Ciais, Josep G. Canadell, Edward J. Dlukokencky, et al. 2016. "The Global Methane Budget 2000–2012." *Earth System Science Data* 8 (2): 697–751.

Saunois, Marielle, Ann R. Stavert, Ben Poulter, Philippe Bousquet, Josep G. Canadell, Robert B. Jackson, and Peter A. Raymond, et al. 2020. "The Global Methane Budget 2000–2017." *Earth System Science Data* 12 (3): 1561–23.

Scarpelli, T., D. Jacob, J. Maasakkers, M. Sulprizio, J.-X. Sheng, K. Rose, L. Romeo, et al. 2019. "A Global Gridded (0.1 X 0.1) Inventory of Methane Emissions from Oil, Gas, and Coal Exploitation Based on National Reports to the United Nations Framework Convention on Climate Change." *Earth System Science Data* 12 (1): 563–75.

Schulz, Rebecca, Christophe McGlade, and Peter Zeniewski. 2020. "Putting Gas Flaring in the Spotlight: New Perspectives on a Persistent Challenge." International Energy Agency, December 9. https://www.iea.org/commentaries/putting-gas-flaring-in-the-spotlight.

Schwietzke, S., O. A. Sherwood, L. Bruhwiler, J. B. Miller, G. Etiope, E. J. Dlugokencky, S. E. Michel, et al. 2016. "Upward Revision of Global Fossil Fuel Methane Emissions Based on Isotope Database." *Nature* 538: 88–91.

Turner, A., D. Jacob, K. Wecht, J. Maasakkers, E. Lundgren, A. E. Andrews, S. C. Biraud, et al. 2015. "Estimating Global and North American Methane Emissions with High Spatial Resolution Using GOSAT Satellite Data." *Atmospheric Chemistry and Physics* 15 (2): 7049–69.

2 Investors Landscape

This chapter analyzes several categories of investors that could finance flaring and methane reduction (FMR) projects. The list is necessarily broad, given the heterogeneity of these projects and limited number of precedents that could translate into standard investment formats.

The chapter also discusses the applicability to FMR projects of new financial instruments to catalyze institutional investor capital, such as green and transition bonds and loans. On the surface, projects aimed at reducing greenhouse gas (GHG) emissions would appear to be prime candidates for the issuance of green bonds and loans. However, these instruments—and the investors that purchase them—apply strict eligibility criteria that, in most cases, rule out the funding of FMR projects. Transition bonds and loans, by contrast, have broader eligibility criteria but are still a niche product.

CATEGORIES OF INVESTORS

The categories of investors analyzed in this section include oil and gas companies (as operators of fields where FMR projects could be implemented), commercial banks, private capital funds (such as private equity and infrastructure), equipment suppliers, petroleum service providers, project developers, development finance institutions, and strategic investment funds (public-sponsored funds that pursue both financial returns and policy goals). For each category, this section provides an estimate of financial firepower and identifies constraints to investors' involvement in FMR projects, including the implications of environmental, social, and corporate governance (ESG) principles and sustainability policies adopted by some of the investor classes.

Oil and gas companies

Operators of flaring fields or other sources of flaring and methane emissions across the value chain should be a natural source of funding for FMR projects. Not only are they the emitters, but they are also the ones that could directly capture any financial returns arising from FMR projects. However, even when the technical and market access barriers discussed in chapter 1 can be overcome, oil and gas companies may still be reluctant to invest in FMR, for two reasons:

1. The overall financial capacity for investments by oil and gas companies varies significantly at different phases of the commodity cycle, with severe drops when oil and gas prices and demand decrease, as in 2015 and most noticeably in 2020. Figure 2.1 shows the industry's global upstream oil and gas capital expenditures ("capex") in the 2010–20 period. After sustained growth in the first half of the decade, coinciding with a benign oil price environment, capex dropped significantly in the second half.

2. Capital expenditures by oil and gas companies tends to be directed toward core activities, such as completion of upstream developments or core storage and transport infrastructure. Apart from a limited number of megaprojects, FMR investments tend to be small and often highly specific to location and field; as a result, oil and gas operators perceive such investments as an unnecessary diversion of scarce engineering and funding resources. In many cases, revenue that can be generated from monetizing a small flare is equivalent to a small fraction of the oil extraction revenue from the same flaring field.

At the same time, commitments by oil and gas companies to reduce scope 1 (direct) and scope 2 (indirect) emissions may lead to increasing investment in FMR. As noted in chapter 1, a few oil companies have already committed to reducing or eliminating routine flaring. In addition to company strategies, regulatory pressure is likely to play a major role in encouraging increased FMR expenditures going forward (see the subsection in chapter 1 titled "Review of Gas Flaring and Venting Regulation").

FIGURE 2.1

Global investments in oil and gas upstream and percentage change from previous year, 2010–20

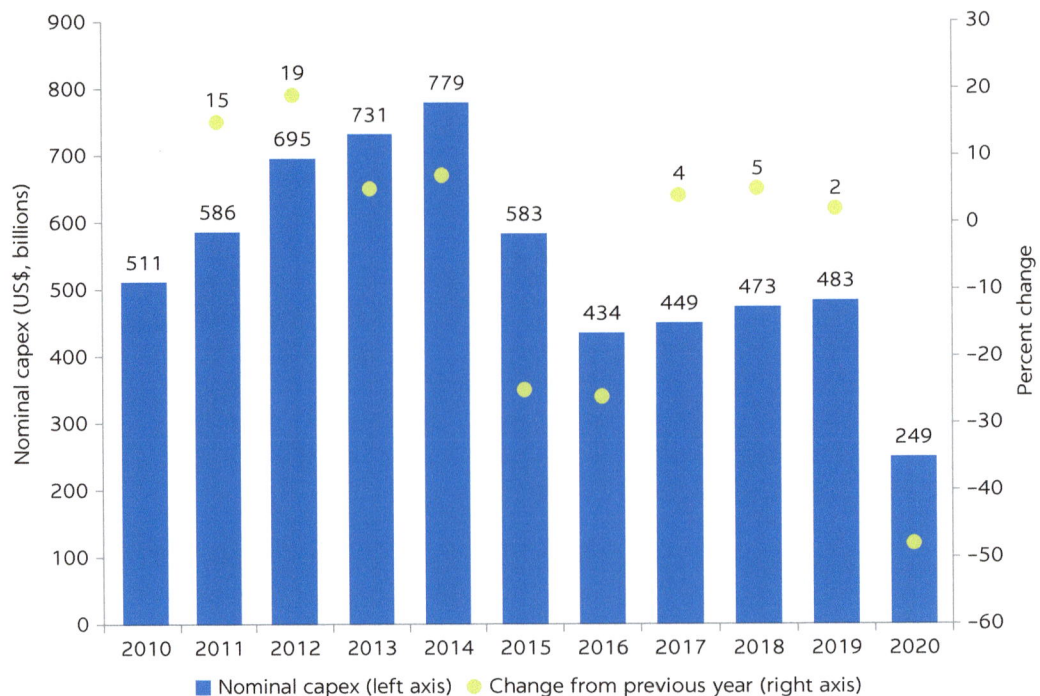

Source: IEA 2020.
Note: Capital expenditures (capex) for 2020 are estimated.

Commercial banks

Despite increasing activism to reduce exposure to the fossil fuel sector, banks continue to be a very significant source of funding for the industry. According to a report by the Rainforest Action Network (2021), the world's 60 largest investment and commercial banks funded the fossil fuel industry to the tune of US$3.8 trillion in the five years following the Paris Agreement (2016–20), by lending and underwriting of debt and equity issuance (figure 2.2).[1] US banks dominate the global league table of fossil fuel finance, but Canadian, Japanese, Chinese, and European banks also figure prominently. About 69 percent of fossil fuel financing analyzed in the report was for oil and gas companies (Rainforest Action Network 2021).

The report also identifies the top-100 fossil fuel companies with expansion plans and finds that they have had no lack of funding in recent years (Rainforest Action Network 2021).[2] The 60 banks analyzed provided US$1.5 trillion to companies with expansion plans in 2016–20. The annual volume of funding for fossil fuels has been steadily increasing since 2016, although it fell by almost 9 percent from 2019 to 2020, after three years of increases of between 4.4 percent and 5.5 percent per year. This evolution over the past four years suggests that fossil fuel sector developments (such as oil price fluctuations) prevail over banks' ESG commitments in driving the supply of finance to the sector. The possible exception is coal mining projects, for which financing from the 60 banks covered by the Rainforest Action Network (2021) increased slightly in 2020, to US$25.4 billion (25 percent higher than in 2016). About 80 percent of the financing for coal mining came from 11 Chinese banks. Financing for coal power—the sector for which many banks have had restrictive policies in place for the longest period—declined by 9 percent in 2020.

The adoption of bank policies restricting financing for fossil fuel expansion or fossil fuel companies altogether is gaining traction, but at a slow pace. The Rainforest Action Network (2021) conducted a point-based evaluation of the policies adopted by the 35 largest lenders to the fossil fuel sector (see box 2.1 on methodology). Most policies address coal, but an increasing number are starting to also address oil and gas—particularly the tar sands and arctic oil and gas segments. European banks are by far the best performing, occupying the first 11 spots in the ranking. American banks are mostly in the middle of the pack, and the bottom half of the ranking is dominated by Asian (Chinese, Japanese, and Singaporean) and Canadian banks. Even at the top of the ranking, however, the score is far from optimum. Italian bank Unicredit, the highest-ranked globally by overall fossil fuel policy, earned only 83.5 points out of 200 available (just over 40 percent of the way toward a full score). In addition, for almost all banks assessed, the main determinant of the overall fossil fuel policy score is commitments to reduce exposure to coal. Furthermore, although many policies focus on project-specific finance, 95 percent of fossil fuel financing is marked as general corporate purposes, and 65 percent of fossil fuel financing in 2020 was through the underwriting of bond and equity issuances, which are often not covered in banks' overall climate policies. When looking at oil and gas policy only, the highest-ranked bank globally is BNP Paribas, with 26.5 points out of 120 (only 25 percent of the way toward a full score, and 4 points lower than its 2019 score). The oil and gas policy ranking follows the same geographic pattern as the overall fossil fuel ranking, with European banks at the top, US banks in the middle, and Asian and Canadian banks at the bottom. See figure 2.3 for the complete oil and gas policy ranking.

FIGURE 2.2

Financing of the fossil fuel industry, by bank, aggregate 2016–20

US$ billions

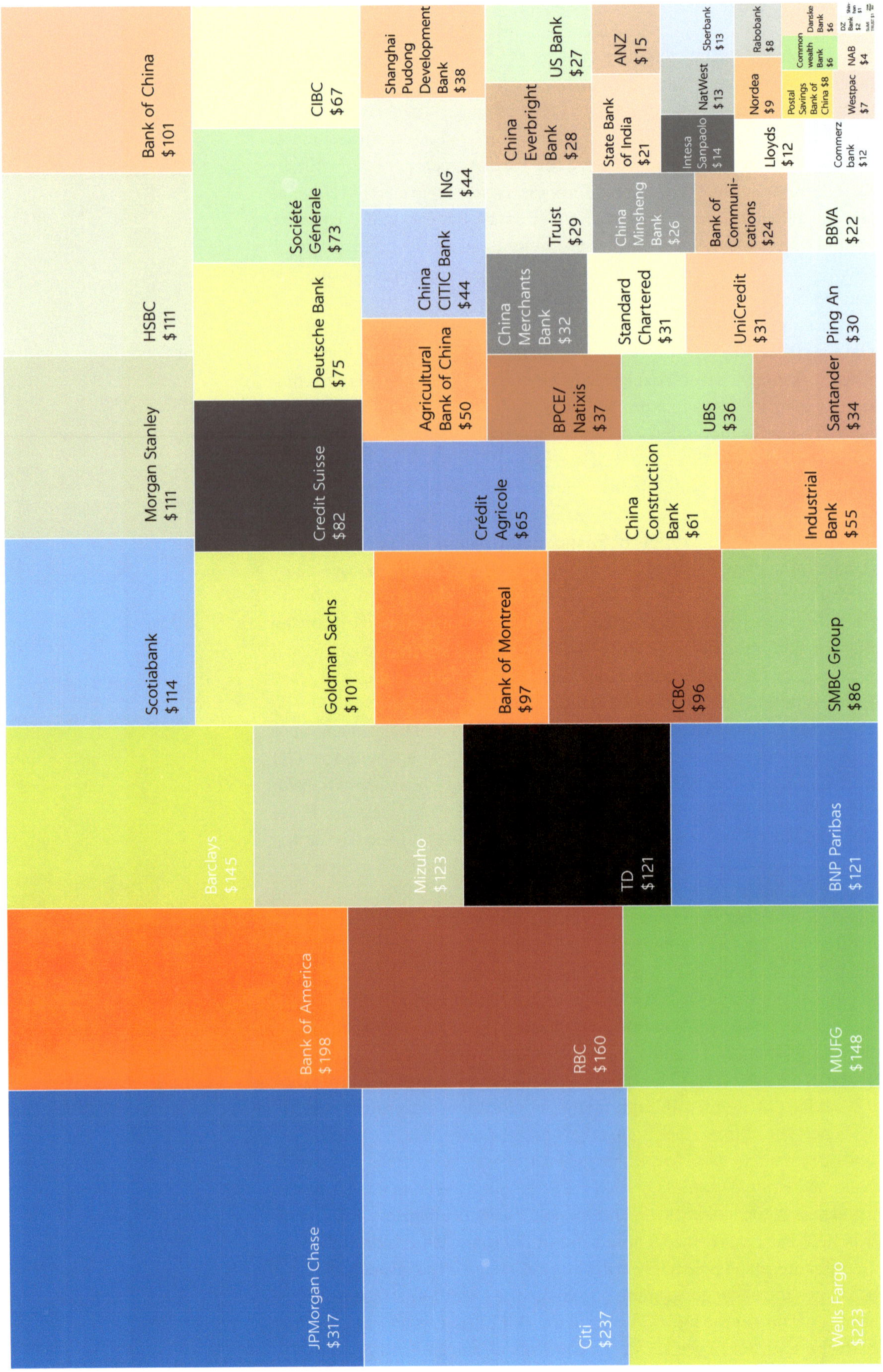

Source: Based on Rainforest Action Network 2021.

Banking on Climate Change 2021's bank policy scoring

Banking on Climate Change 2021 assessed all published bank policies pertaining to the fossil fuel sector and assigned scores. Banks were sent the assessment and invited to comment. Advocacy groups with expertise on the fossil fuel banks based in certain countries were also consulted.

For each fossil fuel and for fossil fuels overall, bank policies were ranked with a point system, based on four parameters:

- Does the bank restrict financing for expansion, via (1) restrictions on direct financing for projects and/or (2) restrictions on financing for expansion companies?
- Does the bank commit to (3) phase out financing for and/or (4) exclude companies active above a certain threshold?

The answer to each of the questions was assigned a decreasing score based on whether a bank expressed a strong, moderate, weak, or no commitment to prohibit or phase out financing, as applicable. Scores were assigned for banks' policy commitments in each of the following sectors: tar sands, arctic oil and gas, offshore oil and gas, fracked oil and gas, liquefied natural gas, other oil and gas, coal mining, and coal power. All oil and gas sector policies in the aggregate can earn a maximum of 120 points. Coal mining and power policies can earn in the aggregate a maximum of 80 points. The maximum overall fossil fuel policy score is 200.

Source: Rainforest Action Network 2021.

Private capital funds focused on natural resources

Private capital funds invest, usually through limited partnership structures, in unlisted companies or assets. They can be further broken down into private equity, venture capital, private debt, real estate, infrastructure, and natural resources funds (Prequin 2020c). Their focus on unlisted investments makes them more suitable for the financing of FMR projects than other asset managers, such as pension funds and insurance companies, that invest primarily in listed securities (such as bonds and stocks).

As of June 2019, private capital funds with a focus on natural resources managed a total of US$765 billion (Prequin 2020a). Of this total, US$230 billion was allocated to funds that invest exclusively in natural resources (for example, oil and gas exploration and production, mining, farmland, and timberland). Infrastructure and other private capital funds that also invest in natural resources account for the remainder (Prequin 2020a). These funds are still relevant as potential investors in FMR projects, which include infrastructure-like investments (for example, construction and operation of pipelines for gas-to-market solutions). In 2019 there were 1,258 active natural resource fund managers; however, after years of significant assets under management growth, 2019 was a challenging year for fundraising. The total amount raised was US$109 billion, but this amount included the launch of two infrastructure megafunds. Excluding those two megafunds, the remaining US$77 billion raised represented a five-year low (Prequin 2020a).[3] Natural resources funds attract capital from a wide range of investors, including foundations, private and public pension funds, endowments, family offices, other asset managers, banks, and insurance companies.

FIGURE 2.3

Total oil and gas policy score of top 60 fossil fuel banks

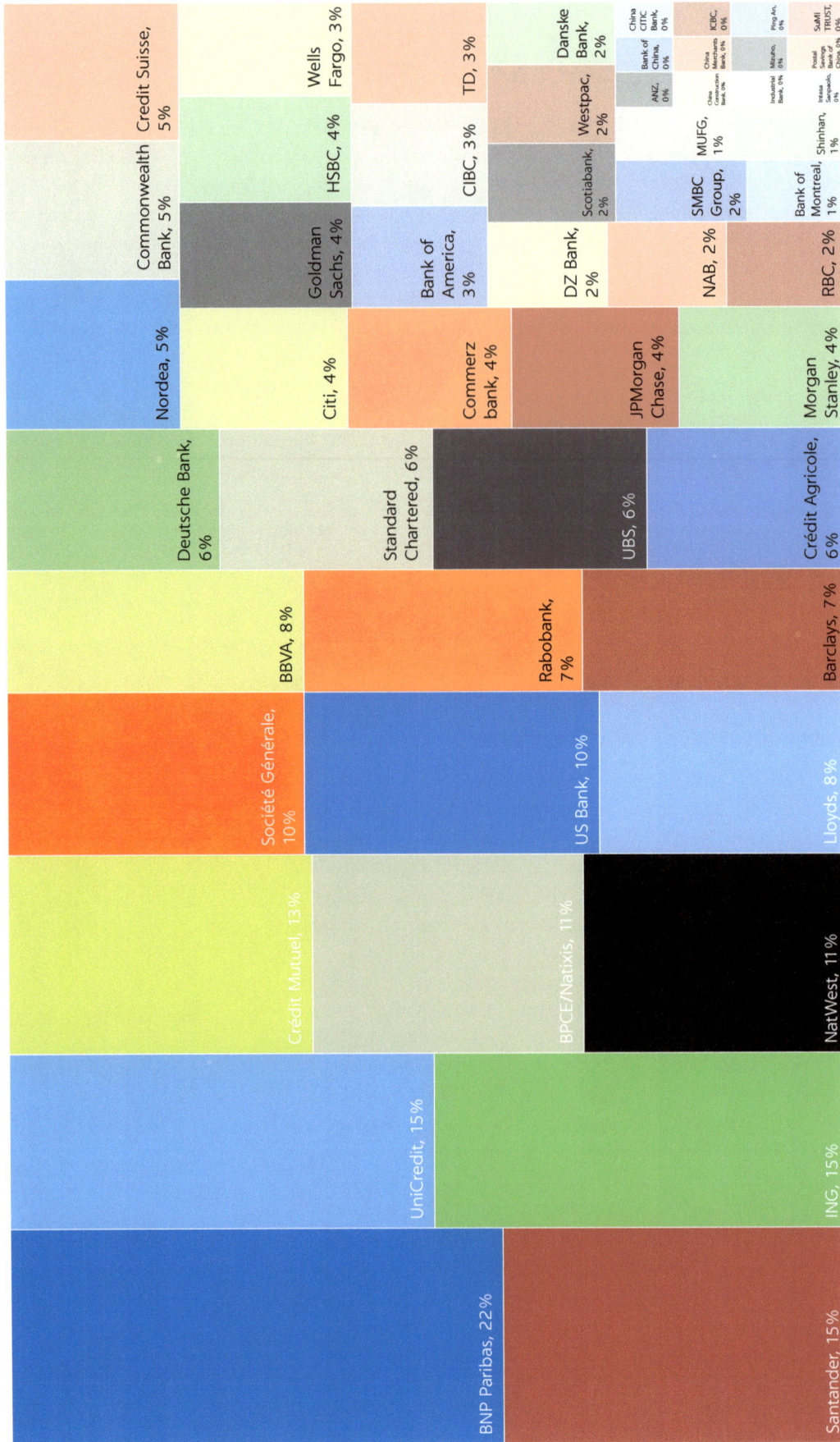

BNP Paribas, 22%

UniCredit, 15%

ING, 15%

Santander, 15%

Crédit Mutuel, 13%

BPCE/Natixis, 11%

NatWest, 11%

Société Générale, 10%

US Bank, 10%

Lloyds, 8%

BBVA, 8%

Rabobank, 7%

Barclays, 7%

Deutsche Bank, 6%

Standard Chartered, 6%

UBS, 6%

Crédit Agricole, 6%

Nordea, 5%

Citi, 4%

Commerzbank, 4%

JPMorgan Chase, 4%

Morgan Stanley, 4%

Commonwealth Bank, 5%

Goldman Sachs, 4%

Bank of America, 3%

DZ Bank, 2%

NAB, 2%

RBC, 2%

Credit Suisse, 5%

Wells Fargo, 3%

HSBC, 4%

CIBC, 3%

TD, 3%

Scotiabank, 2%

Westpac, 2%

Danske Bank, 2%

SMBC Group, 2%

Bank of Montreal, 1%

MUFG, 1%

Shinhan, 1%

Bank of China, 0%

ANZ, 0%

China CITC Bank, 0%

China Merchants Bank, 0%

China Construction Bank, 0%

ICBC, 0%

Mizuho, 0%

Industrial Bank, 0%

Ping An, 0%

Postal Savings Bank of China, 0%

Intesa Sanpaolo, 0%

SuMi TRUST, 0%

Source: Based on Rainforest Action Network 2021.
Note: This chart shows scores as a percentage of 120 (the maximum total oil and gas policy score that could be assigned in the report).

In terms of geographic focus, North America and Europe account for most of the investment, but appetite for funds investing in the rest of the world, especially Africa, has increased recently. North America and Europe represented 57 percent and 34 percent, respectively, of aggregate capital raised in 2010–19. Asia represented 1 percent of capital raised over that period, and the rest of the world (primarily Africa) represented the remaining 8 percent (Preqin 2020b). Africa-focused funds saw a significant increase in fundraising in 2019 (30 percent of funds closure in the rest-of-the-world region and 10 times more capital than in 2017), bucking the global trend for natural resources funds.

The vast majority of natural resources funds target investments in the energy sector, accounting for US$643 billion of assets under management or 84 percent of total funding (Preqin 2020a). Their dry powder—the amount of capital committed by fund investors that is still uninvested—was some US$230 billion as of December 2019.[4] Energy funds invest in the discovery, production, storage, distribution, and retail of energy resources (Preqin 2020c). They specialize in one or more resources, including oil, natural gas, uranium, coal, and renewables. Energy also dominated fundraising in 2019, with a total of US$102 billion raised.[5] Table 2.1 lists the largest funds raised in the energy sector in 2019.

Natural resources funds were slow to embrace sustainable investing, but the trend is accelerating. A survey conducted by Preqin in November 2019 found that only 23 percent of natural resources funds interviewed had an ESG investment policy and that another 11 percent planned to adopt one within the following year—making natural resources funds the second-lowest alternative asset class by ESG implementation (Preqin 2020a).[6] Legacy investments in oil and gas are one possible explanation. However, Preqin's survey also found that 71 percent of investors in alternative funds—of which natural resources funds are a subset—believe that the adoption of ESG policies is or will become integral

TABLE 2.1 **Largest energy funds raised in 2019**

FUND	FIRM	FUND SIZE (US$, BILLIONS)	GEOGRAPHIC FOCUS
Global Infrastructure Partners IV	Global Infrastructure Partners	22.0	North America
EQT Infrastructure IV	EQT	10.1	Europe
Ardian Infrastructure Fund V	Ardian	6.9	Europe
Macquarie European Infrastructure Fund VI	Macquarie Infrastructure and Real Assets (MIRA)	6.7	Europe
North Haven Infrastructure Partners III	Morgan Stanley	5.5	North America
GSO Energy Select Opportunities Fund II	GSO Capital Partners	4.5	North America
NGP Natural Resources XII	NGP Energy Capital Management	4.3	North America
AMP Capital Infrastructure Debt Fund IV	AMP Capital Investors	4.0	North America
European Diversified Infrastructure Fund II	First Sentier Investors	2.9	Europe
Quinbrook Low Carbon Power Fund	Quinbrook Infrastructure Partners	1.6	Rest of world
Actis Long Life Infrastructure Fund	Actis	1.2	Rest of world
Equitix Fund V	Equitix	1.2	Europe
Climate Investor One	Climate Fund Managers	0.9	Rest of world
Ping An Global Infrastructure Funds	Ping An Overseas Holdings	0.8	Rest of world
MIRA Agriculture Fund	Macquarie Infrastructure and Real Assets (MIRA)	0.7	Rest of world

Source: Preqin 2020a.

to the alternative assets sector. This belief is also reflected in the shift of fund-raising away from conventional energy–focused funds to renewables-focused ones; the latter represented 44 percent of the capital raised by energy-focused funds in 2019, compared to just 21 percent in 2010 (Prequin 2020a).

FMR investments are a niche largely unexplored by energy-focused funds at the time of writing this book. Anecdotal evidence, however, points to an increase in interest, consistent with the broader shift toward sustainable investing. For example, OGCI Climate Investments, a US$1 billion fund capitalized by members of the Oil and Gas Climate Initiative whose members account for over 30 percent of global oil and gas production, is also seeking investments in projects or companies that aim to reduce carbon dioxide (CO_2) and especially methane emissions from the oil and gas, industrial, and commercial transport sectors. Although most of its investments at the time of writing are in companies that manufacture and sell emissions-reduction technologies, the fund is actively pursuing project-level FMR investments (OGCI 2020).

Equipment suppliers, petroleum service providers, and project developers

Equipment suppliers and project developers could increasingly become a source of funding for FMR projects. The oil and gas demand shock arising from the COVID-19 (Coronavirus) pandemic has raised additional questions about the potential for oil demand growth in the context of a future low-carbon economy, prompting some engineering, procurement, and contracting (EPC) firms and original equipment manufacturers to target new segments to position themselves for the energy transition, particularly when knowledge synergies can be tapped. Aker and TechnipFMC, for example, have restructured their businesses to create dedicated units for low-carbon projects (Exarheas 2020).

Oil and gas equipment suppliers and EPC firms are increasingly adopting flexible remuneration models that, in effect, position them as investors. Especially in downward phases of the capex cycle, when oil and gas companies refocus on core investments, the ability to provide equipment or EPC services without requiring up-front payment can help suppliers and contractors gain or retain otherwise lost business. The British company Aggreko and the Austrian company Hoerbiger, for instance, offer turnkey power generation solutions using associated gas and selling the electricity to the oilfield operator or to an existing grid. Clients do not pay for the generation and ancillary equipment up front but enter into long-term contracts that allow Aggreko and Hoerbiger to recoup costs and realize a return from the sale of electricity (see the case studies in chapter 4). Galileo, an Argentine manufacturer and supplier of distributed compressed natural gas and liquefied natural gas equipment, also provides turnkey solutions whereby remuneration comes from the sale of the regasified product (see the case study in chapter 4). The British EPC firm Capterio offers a variety of financing structures, including (1) structuring joint ventures through which both Capterio and the client oil company invest and share returns; (2) revenue sharing arrangements, under which Capterio obtains a high share of FMR project revenues until payback of its investment, and a lower share thereafter; (3) accelerated payouts followed by the handover of the FMR facility to the owner of the oil and gas asset; and (4) processing and capacity fees charged to the owner.[7] Appendix A contains a list of service companies actively providing FMR solutions.

Although large original equipment manufacturers and oil and gas service companies are increasingly transforming their portfolios to play an active role in the energy transition, their focus is mainly on new technologies or the deployment of technologies at scale rather than on investing in medium-size flare reduction or providing small turnkey solutions supported by long-term repayment schemes to FMR project developers. Examples include the recently created Schlumberger New Energy, which focuses on low-carbon and carbon-neutral energy technologies. Its activities include ventures in the domains of hydrogen, lithium, carbon capture and sequestration, geothermal power, and geo-energy for heating and cooling buildings.[8] In a similar way, Baker Hughes provides technology solutions to oil companies to tackle scope 1 and scope 2 emissions from oil and gas (including monitoring and detection of emissions, flare optimization programs, and zero-bleed valves), and it invests in energy technology for low-carbon fuels.[9]

Development finance institutions

Development finance institutions (DFIs) are specialized development banks or subsidiaries set up to support private sector development in emerging market and developing economies through equity investments, long-term loans, and guarantees, usually targeting commercial returns.[10] When owned by one government, they are known as bilateral DFIs; examples include CDC (United Kingdom), FMO (Netherlands), and Proparco (France). Multilateral DFIs are private sector arms of international financial institutions established by more than one country and subject to international law; examples include the International Finance Corporation (IFC, part of the World Bank Group), European Investment Bank, and private sector arms of multilateral development banks such as the African Development Bank and Asian Development Bank.

Although subject to more stringent ESG criteria than commercial banks, DFIs are not necessarily prevented from financing firms in the oil and gas sector. Sectoral policies and exposures vary by institution. For instance, at the time of writing this report, the IFC's climate finance policy specifically allows for the financing and provision of advisory services to projects that reduce gas flaring or fugitive methane emissions in existing oil and gas industry installations. However, starting in 2019, the World Bank Group ceased financing upstream oil and gas projects save for in exceptional circumstances when consideration may be given to financing natural gas projects that have a clear benefit in energy access for poor countries and the project is consistent with the country's Paris Agreement commitments. IFC investment is subject to successful project screening under its Anticipated Impact Measurement and Monitoring framework.[11] The European Investment Bank, by contrast, has recently decided to phase out altogether, starting in 2022, its support for oil and gas production and traditional gas infrastructure (networks, storage, and refining facilities). Table 2.2 summarizes the oil and gas policies of the main multilateral development banks, as of the time of writing this report.

The DFIs' track record in FMR projects is fairly limited to date, with the European Bank for Reconstruction and Development (EBRD) being the most active. At the release of EBRD's 2018 Energy Sector Strategy 2019–23, 4 percent of the EBRD's energy investments were in gas flaring reduction (EBRD 2018; see table 2.3). No new FMR project was reported in the EBRD's Project Summary after April 2020.

TABLE 2.2 Selected development financing institution policies on financing the oil and gas sector

INSTITUTION	OFFICIAL SOURCE	OIL AND GAS FINANCING POLICY
African Development Bank (AfDB)	AfDB Group Energy Sector Policy (2011)	• AfDB supports (1) the environmentally and socially sound production, processing, distribution, and export of African hydrocarbons and (2) power generation from oil and gas. • AfDB does not support oil and gas exploration. • No mention of FMR.
Asian Development Bank (ADB)	ADB Energy Policy (2009)[a]	• ADB supports (1) financing natural-gas-based power plants, because of their environmental benefits (vs. coal) and (2) safety and efficiency improvements in the transportation of oil and LNG. • ADB does not support any oil field exploration or oil field development projects. • No mention of FMR.[b]
Asian Infrastructure Investment Bank (AIIB)	AIIB Energy Sector Strategy (2017, revised 2018)	• AIIB supports (1) fossil-fuel–based generation using commercially available least-carbon technology (for example, in many countries, gas-fired power generation); (2) development, rehabilitation, and upgrading of natural gas transportation, storage, and distribution infrastructure, and control of gas leakage, to foster greater use of gas as a transition to a less carbon-intensive energy mix. • No explicit prohibition to support oil and gas exploration. • No mention of flaring, but "control of gas leakage" (above) could be interpreted as reducing methane emissions.
CDC Group (UK)	No official energy policy	No explicit restrictions on oil and gas financing.
European Bank for Reconstruction and Development (EBRD)	EBRD Energy Sector Strategy 2019–23 (2018)	EBRD supports gas investments, provided that they do not displace less carbon-intensive sources, are designed to facilitate energy transition, are subject to assessments that apply a shadow carbon price, and are consistent with EBRD's environmental and social policies. EBRD does not support upstream oil exploration and oil development projects except in exceptional cases where project's proceeds exclusively target the reduction of GHG emissions or flaring from existing producing fields. However, by the end of 2022, the EBRD will stop financing oil and gas upstream projects altogether. The bank will continue to finance selected projects in the midstream and downstream sectors but only where those projects are aligned with, and significantly contribute to, the goals of the Paris Agreement.
European Investment Bank (EIB)	EIB Energy Lending Policy (2019)	Starting in 2022, the EIB will phase out support to (1) the production of oil and natural gas; (2) traditional gas infrastructure (networks, storage, refining facilities); (3) power generation technologies resulting in GHG emissions above 250 gCO$_2$/kWh of electricity generated, averaged over the lifetime for gas-fired power plants seeking to integrate low-carbon fuels; and (4) large-scale heat production infrastructure based on unabated oil, natural gas, coal, or peat. EIB will also stop lending to polluting companies that want to finance low-carbon projects. This would mean, for example, that EIB will no longer finance an oil company's wind energy project. All recipients of EIB loans will be required to draw up decarbonization plans.
Inter-American Development Bank (IDB)	IDB Energy Sector Framework (2018)	No explicit restrictions found on oil and gas financing.
International Finance Corporation (IFC)	AIMM Sector Framework Brief (Oil & Gas) (March 2019); IFC's Definitions and Metrics for Climate-Related Activities (April 2017)	The IFC recognizes that oil and gas will remain important energy sources during the transition to a low-carbon economy. It can provide financing and advisory services to oil and gas companies that (1) increase access to reliable, affordable power; (2) increase fiscal revenues and other income for the government (especially in upstream); (3) have a direct, indirect, or induced impact on GDP and employment; (4) result in potentially significant environmental and social benefits. It also looks at market creation effects in the form of (1) increase in the number of market participants; (2) improvement in sector resilience and quality of supply; (3) increased connectivity of the oil and gas system; (4) adoption of new sustainability and climate mitigation/adaptation technology/processes/practices that can be replicated by other players; and (5) introduction of inclusive business models. IFC projects in the oil and gas sector are screened for financial and development impact considerations, including the organization's Anticipated Impact Measurement and Monitoring (AIMM) framework. Specifically, the IFC can provide financing and advisory services to projects that reduce gas flaring or fugitive methane emissions in existing oil and gas industry installations. Projects are subject to ex ante and ex post verification, also when the IFC does not finance them directly but through financial intermediaries or investment in third-party funds.

(continued next page)

TABLE 2.2, *continued*

INSTITUTION	OFFICIAL SOURCE	OIL AND GAS FINANCING POLICY
Islamic Development Bank (IsDB)	IsDB Energy Sector Policy (2018).	No explicit restrictions on oil and gas financing, but "safety, operational efficiency and sustainability" are taken into account in upstream and downstream projects. Despite a goal to scale up renewable energy, IsDB has funded oil and gas and coal-fired power plants.

Sources: ADB 2009; AfDB 2011; AIIB 2018; EBRD 2018; EIP 2019; IDB 2018; IFC 2017, 2019; IsDB 2018.
Note: FMR = flaring and methane reduction; gCO₂/kWh = grams of carbon dioxide per kilowatt-hour; GHG = greenhouse gas.
a. In May 2021, ADB published a draft energy policy on Supporting Low Carbon Transition in Asia and the Pacific (https://www.adb.org/documents /draft-energy-policy-supporting-low-carbon-transition-asia-and-pacific). The document is undergoing public consultation.
b. ADB commissioned an independent evaluation that concluded that the current energy policy conflicts with ADB's Strategy 2030 (ADB 2020).

TABLE 2.3 Flaring and methane reduction projects, European Bank for Reconstruction and Development, 2014–20

BORROWER	COUNTRY	YEAR	SIZE AND INSTRUMENT	DESCRIPTION
Intergas Central Asia JSC and KazTransGas Aimak JSC	Kazakhstan	2020	€244 million loan	To help the subsidiaries of KazTransGas optimize the balance sheet, continue energy efficiency improvements, and develop a corporate governance action plan. The project description suggests that part of the loan will support the development of the Methane Emissions and Carbon Intensity Reduction Programme.
KazPetrol	Kazakhstan	2018[a]	US$42 million loan	Finance the implementation of AG use technology, including on-site power generation solutions.
Merlon Petroleum	Egypt, Arab Rep.	2015	US$43 million loan	Finance operations in Egypt, including investments to reduce AG flaring, and capital expenditures for field development.
Pico Oil and Gas	Egypt, Arab Rep.	2015	Up to US$100 million loan	The project has several components, one of which is the demonstration of new replicable activities by implementing a flaring reduction policy with best practices and new technologies.
Kuwait Energy	Egypt, Arab Rep.	2016[a]	US$100 million loan	The project was supposed to support the development of Kuwait Energy's Egyptian operations, including investments in AG flaring prevention and recovery.
Falcon	Kazakhstan	2014[a]	US$20 million loan	Fund capital expenditures and, in particular, the construction of an AG facility to eliminate all routine flaring at the Shoba oil field.

Source: European Bank for Reconstruction and Development, Project Summary Documents (https://www.ebrd.com/work-with-us/project-finance /project-summary-documents.html).
Note: Information provided reflects publicly disclosed data and may not be an exhaustive list of projects financed or to be financed by the European Bank for Reconstruction and Development. AG = associated gas.
a. Transaction was announced but not closed.

Going forward, the criteria for the financing of oil and gas projects, including FMR, may evolve—and possibly tighten—as development banks and DFIs review previously established common standards in this respect. In 2015, as a voluntary joint initiative, the members of the Multilateral Development Banks' Climate Finance Tracking Working Group and the International Development Finance Club Climate Finance Working Group agreed on a set of Common Principles for Climate Mitigation Finance Tracking (World Bank 2015a).[12] These principles were updated in October 2021 (EIB 2021) to reflect an enhanced ambition for reducing GHG emissions, taking into account new mitigation activities required to meet the goals of the Paris Agreement and excluding activities that, although reducing GHG emissions in the short term, may maintain highly emissive technologies for several more years, thereby undermining the long-term temperature goal. The use of associated natural gas from brownfields that would otherwise be flared as a feedstock or fuel to supply electricity, heat, mechanical energy, or cooling, when such use decreases GHG emissions substantially, as well as the reduction of fugitive GHG emissions in existing energy transportation or storage infrastructure are among the eligible activities, suggesting a possible space for FMR projects.

In August 2021 the US Treasury published a Guidance on Fossil Fuel Energy of the Multilateral Development Banks requiring the United States to oppose fossil fuel projects at the World Bank Group, Asian Development Bank, African Development Bank, and InterAmerican Development Bank, except in extremely rare circumstances outlined in the document (US Treasury 2021). The guidance is designed to accelerate the transition to more sustainable, climate-smart economies and will affect capital allocation by multilateral development banks independently of their individual oil and gas financing policies.

Strategic investment funds

Strategic investment funds (SIFs) are special purpose investment vehicles, backed by governments or other public institutions, that pursue both financial returns and policy objectives and aim to mobilize commercial capital for investments otherwise avoided by private investors (Halland, Noel, and Tordo 2016). Policy objectives vary by SIF and can include accelerating a country's economic development, employment creation, climate change mitigation and adaptation, and infrastructure development. Sometimes SIFs are part of sovereign wealth funds. Some 30 SIFs have been established since 2000 at the national level, with a significant acceleration after the global financial crisis. In addition, many SIFs were established at the subnational government level. When considering the wider universe of multilateral and national SIFs, 20 of them include oil and gas within their investment mandate, but only 16 among the 20 have invested in the sector (table 2.4). SIF reporting is generally not detailed, and disaggregated portfolio allocation is not commonly disclosed.

SIFs could be meaningful contributors of capital to FMR projects and help fill the gap left by other funding sources. SIFs enjoy a close understanding of a country's investment environment, including access to its project pipeline. Some SIFs play the role of project developers—not just passive investors—and could facilitate FMR investments that in most cases are not off the shelf. SIFs, especially those domiciled in oil and gas economies, are also likely to have no or few restrictions with regard to investing in the fossil fuel sector (Halland, Noel, and Tordo 2016). For example, the US$650 million Nigeria Infrastructure Fund, an SIF established and managed by the Nigeria Sovereign Investment Authority, is indirectly funded with a portion of the country's hydrocarbon revenues and has the ability to invest in gas pipeline, infrastructure, and storage projects and the broad power sector.[13]

FINANCING INSTRUMENTS

The asset management community has developed new financial products to cater to the desire of pension funds, insurance companies, and other institutional fund managers (including endowment funds, foundations, and family offices) to allocate capital to investments that not only produce attractive risk-adjusted returns but also contribute to climate change adaptation and mitigation. Green bonds and, to a much lesser extent, green loans are the most established of these new asset classes. However, the eligibility of FMR projects for these instruments is questionable, according to increasingly well-established industry standards.

TABLE 2.4 **Strategic investment funds that invest in oil and gas**

MANDATE	FUND	NATIONAL OR MULTILATERAL	GEOGRAPHIC FOCUS
Oil and gas in mandate and disclosed project in the sector	Silk Road Fund	National	China/global
	Partnership Fund	National	Georgia
	Ghana Infrastructure Investment Fund	National	Ghana
	Palestine Investment Fund	National	Palestine
	Russian Direct Investment Fund	National	Russian Federation / global
	Public Investment Fund	National	Saudi Arabia / global
	Turkiye Wealth Fund	National	Turkey
	Mubadala Investment Company	National	Abu Dhabi / global
	Investment Corporation of Dubai	National	Dubai/global
	IFC Africa, Latin America and Caribbean Fund	Multilateral	Africa and Latin America and the Caribbean
	IFC China-Mexico Fund	Multilateral	Latin America and the Caribbean
	IFC Global Infrastructure Fund	Multilateral	Global emerging markets
	IFC Middle East and North Africa Fund	Multilateral	Middle East and North Africa
	Infraco Asia	Multilateral	Asia (South and Southeast)
	Marguerite II	Multilateral	EU and accession countries
	Philippine Investment Alliance for Infrastructure	Multilateral	Philippines
Oil and gas in mandate	Nigerian Sovereign Investment Authority (National Infrastructure Fund)	National	Nigeria
	Africa50	Multilateral	Africa
	Emerging Africa Infrastructure Fund	Multilateral	Sub-Saharan Africa
	Marguerite I	Multilateral	EU and accession countries

Source: World Bank, based on publicly available information.
Note: This list may not be exhaustive because many strategic investment funds do not disclose their investments or strategies. EU = European Union; IFC = International Finance Corporation.

Transition bonds and loans are meant to bridge the gap, financing emitting activities on a credible path to decarbonization. Although still niche products, they may be more suitable to FMR financing. The following subsections discuss green and transition bond and loan categories.

Green bonds

Green bonds are debt securities issued to raise capital specifically to support climate-related or environmental projects (World Bank 2015b). They are the most popular debt instrument with environmental aspirations; global issuance reached US$271 billion in 2019 (*The Economist* 2020). When deciding whether to buy a green bond, investors assess both its financial terms (for example, maturity, interest rate, and credit quality of the issuer) and the environmental purpose of the project(s) that the bond is intended to fund. The benefit for the bond issuer is the ability to reach the expanding universe of impact-oriented institutional investors (including pension funds and insurance companies) and showcase its environmental credentials, while paying interest rates generally comparable to those of its conventional bonds (*The Economist* 2020).

In principle, nothing prevents the use of green bonds to fund FMR projects. Green bonds are not subject to mandatory standards and definitions. Provided that issuers and investors voluntarily agree on the climate-related or environ-mental purpose of a bond, it can be classified as green. This practice has led to the blurring of boundaries between sustainable and nonsustainable projects funded by green bonds (so-called greenwashing). For instance, in China green bonds have been issued to finance clean coal projects—a use of proceeds that would not comply with the sustainability policies of many European and North American institutional investors.[14]

In practice, backlash against greenwashing has resulted in the increasing adoption of green bond standards with tight eligibility criteria that in most cases rule out the use of proceeds for FMR projects. The Green Bond Principles, pub-lished by the International Capital Markets Association, are the best-known set of voluntary guidelines for the issuance of green bonds (ICMA 2018). The Green Bond Principles establish guidelines on the process for project evaluation and selection, management of proceeds, and reporting (see box 2.2 for a summary of

---

BOX 2.2

The European Union Green Bond Principles: Overview

The Green Bond Principles (GBP) are voluntary pro-cess guidelines that clarify the approach for the issu-ance of green bonds to promote transparency and integrity in the development of this market. The GBP's target audience includes prospective issuers who need guidance on the launch of green bonds, investors who need to evaluate the environmental impact of their green bond investments, and underwriters who bene-fit from recognized standards when marketing trans-actions to investors.

The GBP have four components:

1. *Use of proceeds*. The GBP list several broad eligi-bility categories for the issuance of green bonds: renewable energy, energy efficiency, pollution prevention and control, environmentally sustain-able management of living natural resources and land use, terrestrial and aquatic biodiversity con-servation, clean transportation, sustainable water and wastewater management, climate change adaptation, eco-efficient and/or circular econ-omy–adapted products, production technologies and processes, and green buildings.

2. *Process for project evaluation and selection*. Issuers are recommended to clearly communicate to investors the following: the environmen-tal sustainability objective, the process used to determine the project's fit with eligibility categories, and any eligibility or exclusion criteria

as well as process for the identification and miti-gation of environmental and social risks.

3. *Management of proceeds*. The GBP recommend that issuers create dedicated subaccounts and formal internal processes to track lending and investment operations linked to green bonds. The principles also encourage the appointment of auditors or other third parties to verify the track-ing methods and allocation of proceeds.

4. *Reporting*. The GBP recommend the preparation of annual reports (until full allocation of the pro-ceeds) including a list of the projects funded with green bond proceeds, a brief description of the projects, the amounts allocated, and the projects' expected impact. If possible, issuers should com-plement qualitative information with quantitative performance data, as well as disclosure of impact assessment methodology and assumptions.

The GBP also recommend that green bond issuers appoint independent external reviewers to confirm the alignment of the bond with the four GBP components. The content of external reviews may vary in scope, and could include second-party opinions on the alignment with the GBP, independent verification against a speci-fied set of criteria, certification against a recognized green standard label, or assessment of the green bond or specific features (for example, use of proceeds) based on an established scoring or rating methodology.

Source: ICMA 2018.

the Green Bond Principles). They also recognize several broad categories of eligibility (none of which is in the fossil fuel industry) but do not provide prescriptive views on specific technologies and solutions. Several international and national initiatives have tried to fill the gap by producing detailed taxonomies of eligible green-bond-funded projects. Examples include the following:

- The Climate Bonds Initiative (CBI), an international nonprofit organization dedicated to mobilizing the bond market for climate change solutions, has published one of the most prominent taxonomies, which it uses to label bond issuances as "Climate Bond Certified" (in the same spirit as the Fair Trade certificate). The CBI taxonomy excludes any uses of proceeds that foster the continuation of the fossil fuel industry, including investment in FMR projects (CBI 2020a). The CBI certification is rapidly gaining traction, with a cumulative US$150 billion in certified bond issuance as of 2020, amounting to about 14 percent of the US$1.1 trillion green bond market in the same year (CBI 2021).

- The European Union's Technical Expert Group on Sustainable Finance has recently proposed a voluntary European Union Green Bond Standard (EU GBS) to enhance the effectiveness, transparency, comparability, and credibility of the green bond market and to encourage market participants to issue and invest in EU green bonds.[15] The eligibility list for EU green bonds would be aligned with the broader EU Taxonomy Regulation for environmentally sustainable economic activities (European Union 2020). Although it does not explicitly mention FMR in one way or another, the taxonomy contains wording that points to challenges in considering FMR a sustainable activity.[16] EU GBS could be implemented on a voluntary basis by both EU and non-EU green bond issuers, and therefore have quite a broad reach (European Commission 2020). The European Commission is currently considering the possibility of a legislative adoption of the EU GBS.

Green loans

Green loans are loan instruments used exclusively to finance or refinance, totally or partially, new or existing eligible green projects (Loan Market Association 2018). Green loans mirror green bonds in aim and application but are a much less established asset class. CBI (2020c) estimates that only US$10 billion worth of green loans was underwritten in 2019. The Loan Market Association (2018) published a set of Green Loan Principles that follow very closely the structure of the International Capital Markets Association's Green Bond Principles and list the same broad eligibility categories for use of proceeds.

FMR projects are likely to face eligibility issues for green loans and green bonds. The CBI taxonomy, for instance, applies not just to bonds but to a variety of debt instruments, including loan facilities and syndicated loans.[17]

Transition bonds and loans

FMR projects could be eligible for a transition bond or loan—an emerging debt instrument that is meant to fund activities that are not yet operating at or near zero emissions but that present significant decarbonization potential. In response to the tight eligibility criteria of green bonds and loans, several market players have proposed standards for this new category of bonds and loans. All guidelines and proposals are very recent, and none specifically mentions the eligibility of

FMR projects. However, they all share the goal of reducing emissions in high-emitting sectors, opening the door—at least in principle—to FMR applications. Proponents of transition bonds and loans include French asset manager AXA, CBI (in partnership with Credit Suisse), DBS Bank (Singapore), and EBRD. Box 2.3 provides a summary of their transition bond guidelines.

At the time of writing, transition bond issuance has been minimal, especially when compared to the size of the green bond market. Table 2.5 provides a sample of transactions labeled or contextualized by the issuers as transition bonds.

Sustainability-linked loans

Sustainability-linked loans (SLLs) are a type of transition loan. SLLs can take different forms, all less constraining than green loans. In SLLs the terms of the loan are linked to performance under sustainability criteria or third-party assessments of green criteria. This practice has the advantage (compared to green bonds or loans) that loan proceeds can be applied to general corporate purposes, rather than being tied up in a controlled bank account to be applied only for a specific limited green purpose.

BOX 2.3

Transition bond guidelines: Summary

French asset manager **AXA**'s guidelines mirror the four components of the Green Bond Principles (see main text) but include in the use of proceeds non-renewable energy, transportation, and industry, among others. They also recommend that issuers clearly communicate how climate transition fits in their business models and strategies, and that transition strategies should be intentional, material, and measurable (Takatsuki and Foll 2019).

The **Climate Bonds Initiative**'s white paper on transition bonds proposes the use of a "transition" label for eligible investments that (1) make a substantial contribution to halving global emissions by 2030 and reaching net zero emission by 2050 but without playing a long-term role or (2) will play a long-term role, but it is uncertain whether they will be aligned with net zero goals in the long term (CBI 2020c).

DBS Bank has published a framework and taxonomy to label bonds, loans, and other financial instruments as green, aligned with the United Nations Sustainable Development Goals, or transition. The latter would apply to the funding of activities that reduce greenhouse gas emissions more than industry norms or enable the application of less carbon-intensive options. DBS Bank recognizes that assessing transition activities is complicated and must consider contextual information such as location of the activity, best available technology, time horizon, and speed of change toward net zero emissions (DBS Bank 2020).

The **European Bank for Reconstruction and Development** has published criteria for its own issuance of "green transition bonds." The use of proceeds is to finance or refinance "projects that are intended to enable significant improvements towards decarbonization, reduction in environmental footprint and/or improved resource efficiency in key sectors of the economy" (EBRD 2019). At least 50 percent of the proceeds must be specifically used to ensure the green transition of a project or asset; in addition, the related company must commit to ensure improved climate governance across the organization. Examples of eligible sectors include chemical, cement, and steel production; agribusiness (promoting energy efficiency and sustainable land use); and activities that enable a green transition, such as electricity grids, supply chains, low-carbon transport and infrastructure, logistics and information technology, and construction and renovation of buildings (EBRD 2019).

TABLE 2.5 **Sample of transition bonds issued**

ISSUER	YEAR	SIZE	ISSUER'S LABEL	SECTOR	USE OF PROCEEDS
Cadent (United Kingdom)	2020	€500 million	Transition bond	Gas distribution	Bond issued by UK gas distribution company. Use of proceeds includes methane leakage control, network repairs and hydrogen readying, and low-carbon vehicles.
Credit Agricole (France)	2019	€100 million (purchased by AXA)	Transition bond	Various	Refinance existing Credit Agricole commercial loans whose use of proceeds include (1) development of gas-fired power stations in Asia to reduce dependence on goal and oil-fueled electricity generation, (2) loans to shipping companies switching from heavy marine diesel oil to liquid natural gas propulsion, and (3) loans to South American industrial groups implementing energy efficiency and wastewater treatment measures.
European Bank for Reconstruction and Development	2019	€500 million	Green transition bond	Various	Financing investments in sectors that are dependent on the use of fossil fuels, such as manufacturing, food production, and the construction and renovation of buildings. Green transition projects finance improvements in resource usage, for instance by lowering the carbon intensity through energy efficiency measures or through the replacement of a high-carbon asset by a lower-carbon asset.
Snam (Italy)	2019	€500 million	Climate action bond[a]	Gas distribution	Upgrading assets to reduce methane and CO_2 emissions, renewable energy projects, investing in new energy transition businesses such as bio-CNG and small-scale LNG, and energy efficiency services for residential and commercial buildings.

Sources: Based on AXA website, accessed on May 18, 2021 (https://www.axa.com/en/magazine/forming-a-bond-supporting-the-energy-transition0); CBI 2020c.
Note: CNG = compressed natural gas; CO_2 = carbon dioxide; LNG = liquefied natural gas.
a. Although not using the "transition bond" label explicitly, Snam justified this issuance in the context of its commitment to the energy transition (gas as an interim fuel while renewables gain ground).

SLLs are usually structured as revolving facilities with a small incremental pricing benefit to the borrower for meeting agreed-on sustainability targets. For example, at the end of 2019, Shell signed a US$10 billion sustainable revolving credit facility—another type of sustainability loan—whereby interest and commitment fees paid on the facility are linked to Shell's progress toward reaching its short-term net carbon footprint intensity target, as published in its Sustainability Report (Loan Market Association 2021). Sustainable revolving credit facilities are becoming quite popular, particularly in the mining sector and energy-intensive industries. Green reserve-based loans are SLLs collateralized by the borrower's oil and gas reserves and based on the producer's emissions-reduction profile. Although sustainable revolving credit facilities and green reserve-based loans may offer oil companies and oil service companies access to finance to support the transition to low-carbon operations, the setup and monitoring cost may be too high for all but large gas flaring and methane reduction projects.

SUMMARY ASSESSMENT OF FINANCING SOURCES AND INSTRUMENTS

Various sources of capital are potentially available for FMR financing, depending on the size and complexity of the project. Cyclical and technical constraints are important factors in structuring financial solutions for FMR projects. Oil and gas price and demand fluctuations have historically affected investors' appetite. In the current low capex environment, some equipment suppliers and project developers have started to shift to service models that do not require oil and gas companies to shoulder the up-front cost of an FMR project. For all categories of capital providers, the transition toward a low- carbon future is becoming an ever more important factor driving sustainable investment and capital allocation. It should create an incentive for all providers to consider FMR investments, which could also be eligible for the issuance of new securities such as transition loans and SLLs. Sources and instruments for FMR financing are summarized in table 2.6.

TABLE 2.6 Summary assessment of sources and instruments for flaring and methane reduction financing

FUNDING SOURCE OR INSTRUMENT	MARKET SIZE CONSIDERATIONS	APPLICABILITY TO FMR
Oil and gas companies	A large pool of capital, but very sensitive to oil price and demand fluctuations. Spending plans collapsed with the COVID-19 (Coronavirus) crisis. Major companies are restructuring their portfolios toward low-carbon fuels and technology solutions.	Except for very large projects, FMR projects have traditionally been treated as noncore by oil and gas companies. Small projects—in capex and revenue potential—coupled with complex and highly location-specific design make FMR a "nuisance" for many oil and gas operators. However, commitments to reducing oil and gas carbon footprint by many oil companies could direct more capital toward FMR.
Commercial banks	A large pool of capital. Despite pressure to reduce exposure to oil and gas for ESG reasons, the sector remains core for many lending institutions.	Very likely. Banks' commitments to cut lending to oil and gas have so far affected only niche segments such as Arctic oil and tar sands (if any). FMR lending could be considered a step toward greening of the banks' oil and gas exposures. Small project size, however, may be an issue, especially for international banks.
Private capital funds	A large pool of capital, especially in the form of equity.	Questionable for small FMR projects that require significant project development expertise and cost (local funds may have an advantage over large international funds in this segment). More likely for megaprojects for which private capital funds, even large international ones, could fill the equity gap.
Equipment suppliers and project developers	Hard to quantify the size of investment to date, but there is an increasing trend of offering turnkey solutions that do not require up-front capex from oil and gas companies.	Very likely, especially in gas-to-power and for small, medium, and large flares (not megaflares).
Development finance institutions	Not as large as commercial banks or private capital funds, but still a significant source of capital (both equity and debt).	Likely source, especially in EMDEs where DFIs have presence on the ground, and could provide both equity and debt. However, any tightening of the current sustainable finance and ESG policies may result in DFIs reducing exposure to the oil and gas sector.
Strategic investment funds	Fast-growing, but not all countries with FMR projects may have an established SIF.	Very likely, especially for SIFs sponsored and funded by fossil-fuel-rich countries that are looking to decarbonize their economies by shifting to lower-carbon fuels and reducing harmful emissions. High degree of local knowledge and visibility over project pipelines. Ability to take a hands-on approach in project development.
Green bonds/loans	Large and growing asset class.	Limited. Prominent initiatives to establish generally accepted green bond standards are likely to result in the exclusion of FMR from the eligible uses of proceeds.
Transition bonds/loans	Market still in its infancy, but market players are advocating for standardization of criteria to facilitate widespread issuance.	Likely. Although transition bond/loan standards are still in the early stages of development, the rationale for these products—to help high emitters decarbonize—is very much in line with the climate mitigation goal of FMR projects.
Sustainability-linked loans	Growing in popularity.	Likely. Sustainability-linked loans can be structured around the decarbonization targets of the recipient. They are not project specific, allowing for increased ability to balance portfolio strategies. The rationale for these products—to help high emitters decarbonize—is very much in line with the climate mitigation goal of FMR projects. However, set-up and monitoring cost may be too high for all but large gas flaring and methane reduction projects.

Source: World Bank.
Note: capex = capital expenditures; DFI = development finance institution; EMDEs = emerging market and developing economies; ESG = environmental, social, and corporate governance; FMR = flaring and methane reduction; SIF = strategic investment fund.

NOTES

1. The universe analyzed in the report includes 2,300 companies that received funding from the 60 largest banks across the following sectors globally: extraction, transportation, transmission, combustion or storage of any fossil fuels, and fossil fuel–based electricity generation, including coal mining and coal power generation.

2. The 100 expansion companies examined include 60 oil and gas companies, 15 companies behind key pipeline and liquefied natural gas terminals that would enable higher upstream production, and 11 coal mining companies and 15 coal power companies (Rainforest Action Network 2021, 31).

3. The US$77 billion excludes the launch of two infrastructure megafunds (US-based Global Infrastructure Partners IV with US$22 billion raised and Sweden-based EQT Infrastructure IV with US$9 billion raised).

4. World Bank estimate based on figure 2.6 of Preqin (2020a).

5. This amount includes the two previously mentioned infrastructure megafunds, which have an energy focus.

6. Alternative funds are a broader category that includes, as a subset, private capital funds. Hedge funds also belong to the alternative universe.

7. Capterio website (https://capterio.com/our-solution).

8. See the company's website for details on its recent investment (https://newenergy.slb.com /new-energy-sectors).

9. See the company's website for details on its business focus (https://www.bakerhughes .com/energy-transition).

10. Definition from Organisation for Economic Co-operation and Development website (https://www.oecd.org/development/development-finance-institutions-private -sector-development.htm), accessed on May 18, 2021.

11. See Anticipated Impact Measuring and Monitoring framework for a list of criteria considered (https://www.ifc.org/wps/wcm/connect/98183cc6-fbf0-4cad-93cf-18c8ae2c7f66/AIMM-Oil-and-Gas-Consultation.pdf?MOD=AJPERES&CVID=nmTfdsD).

12. The Climate Finance Tracking Working Group comprises the African Development Bank, Asian Development Bank, EBRD, European Investment Bank, Inter-American Development Bank Group (IDBG), and Islamic Development Bank; and the IFC, World Bank (International Development Association and International Bank for Reconstruction and Development), and Multilateral Investment Guarantee Agency of the World Bank Group.

13. Based on information retrieved from the Nigeria Sovereign Investment Authority's website (https://nsia.com.ng/), last accessed on June 3, 2021.

14. In May 2020 the People's Bank of China announced a proposal to exclude the clean use of fossil fuels from the list of projects that may benefit from green finance (Yamaguchi and Ahmad 2020).

15. European Commission website (https://ec.europa.eu/info/business-economy-euro /banking-and-finance/sustainable-finance/eu-green-bond-standard_en).

16. Art. 41 of the EU Taxonomy Regulation excludes from the activities "contributing substantially to climate change mitigation" those that would "hamper the development and deployment of low-carbon alternatives" and "lead to a lock-in of assets incompatible with the objective of climate-neutrality, considering the economic lifetime of those assets." Detailed technical screening criteria for the application of regulation may further clarify the regulation's approach toward fossil fuel investments, including FMR. The publication of such criteria, originally expected by December 2020, has been delayed because of the large number of comments received during the consultation period. The European Commission recognizes "the need to give reassurance that the taxonomy will not block access to finance for enterprises and sectors in transition towards our climate targets." To this end, the Platform on Sustainable Finance, which has replaced the Technical Expert Group, has been tasked to advise it on how the taxonomy could be used to finance transition activities. Details on the platform's role and timeline for deliverables can be found on the European Commission web page (https://ec.europa.eu/info/business-economy-euro /banking-and-finance/sustainable-finance/overview-sustainable-finance/platform -sustainable-finance_en#what).

17. For more information, see CBI website (https://www.climatebonds.net/certification /eligible-instruments).

REFERENCES

ADB (Asian Development Bank). 2009. "Energy Policy 2009." ADB Policy Paper, June, ADB, Mandaluyong, Philippines. https://www.adb.org/documents/energy-policy.

ADB (Asian Development Bank). 2020. "ADB Energy Policy and Program, 2009–2019." Sector-wide evaluation, ADB, Mandaluyong, Philippines. https://www.adb.org/sites/default/files/evaluation-document/518686/files/swe-energy-policy-and-program.pdf.

AfDB (African Development Bank). 2011. "Energy Sector Policy of the AfDB Group." AfDB, Abidjan. https://www.afdb.org/en/documents/document/afdb-group-energy-sector-policy-30043.

AIIB (Asian Infrastructure Investment Bank). 2018. "Energy Sector Strategy: Sustainable Energy for Asia." AIIB, Beijing. https://www.aiib.org/en/policies-strategies/strategies/sustainable-energy-asia/index.html.

CBI (Climate Bonds Initiative). 2020a. 'Climate Bonds Taxonomy." CBI. https://www.climatebonds.net/files/files/CBI_Taxonomy_Tables_January_20.pdf.

CBI (Climate Bonds Initiative). 2020b. "2019 Green Bond Market Summary." CBI. https://www.climatebonds.net/files/reports/2019_annual_highlights-final.pdf.

CBI (Climate Bonds Initiative). 2020c. "Financing Credible Transitions. How to Ensure the Transition Label has Impact." Climate Bonds White Paper, CBI. https://www.climatebonds.net/files/reports/cbi_fincredtransitions_final.pdf.

CBI (Climate Bonds Initiative). 2021. "Sustainable Debt: Global State of the Market." CBI. https://www.climatebonds.net/files/reports/cbi_sd_sotm_2020_04d.pdf.

DBS Bank. 2020. "Sustainable and Transition Finance Framework and Taxonomy." DBS Bank, Singapore. https://www.dbs.com/iwov-resources/images/sustainability/responsible-banking/IBG%20Sustainable%20%26Transition%20Finance%20Framework_Jun2020.pdf?pid=DBS-Bank-IBG-Sustainable-Transition-Finance-Framework-Taxonomy.

EBRD (European Bank for Reconstruction and Development). 2018. "Energy Sector Strategy 2019–2023." EBRD, London. https://www.ebrd.com/power-and-energy/ebrd-energy-sector-strategy.pdf.

EBRD (European Bank for Reconstruction and Development). 2019. "Green Transition Bond / Green Bond Programme Information Template." EBRD, London. https://www.ebrd.com/documents/treasury/framework-for-green-transition-bonds.pdf?blobnocache=true.

Economist. 2020. "What Is the Point of Green Bonds?" The Economist, September 19. https://www.economist.com/finance-and-economics/2020/09/19/what-is-the-point-of-green-bonds.

EIB (European Investment Bank). 2019. "EIB Energy Lending Policy: Supporting the Energy Transformation." EIB, Luxembourg. https://www.eib.org/en/publications/eib-energy-lending-policy.

EIB (European Investment Bank). 2021. "Common Principles for Climate Mitigation Finance Tracking." Version 3, October 18. EIB, Luxembourg. https://www.eib.org/attachments/documents/mdb_idfc_mitigation_common_principles_en.pdf.

European Commission. 2020. "Usability Guide: EU Green Bond Standard." European Commission, Brussels. https://ec.europa.eu/info/sites/info/files/business_economy_euro/banking_and_finance/documents/200309-sustainable-finance-teg-green-bond-standard-usability-guide_en.pdf.

European Union 2020. "Regulation (EU) 2020/852 of the European Parliament and of the Council of 18 June 2020 on the Establishment of a Framework to Facilitate Sustainable Investment." Official Journal of the European Union, June 22. https://eur-lex.europa.eu/legal-content/EN/TXT/PDF/?uri=CELEX:32020R0852&from=EN.

Exarheas, Andreas. 2020. "Major Oil EPC Players Shifting toward Cleaner Energy." Rigzone, November. https://www.rigzone.com/news/major_oil_epc_players_shifting_toward_cleaner_energy-02-nov-2020-163728-article/.

Halland, Havard, Michel Noel, and Silvana Tordo. 2016. "Strategic Investment Funds: Opportunities and Challenges." Policy Research Working Paper 7851, World Bank, Washington, DC. https://openknowledge.worldbank.org/handle/10986/25168.

ICMA (International Capital Markets Association). 2018. "Green Bond Principles: Voluntary Process Guidelines for Issuing Green Bonds." ICMA. https://www.icmagroup.org/assets/documents/Regulatory/Green-Bonds/Green-Bonds-Principles-June-2018-270520.pdf.

IDB (Inter-American Development Bank). 2018. "Energy Sector Framework Document: Energy Division." IDB, Washington, DC. https://www.iadb.org/en/sector/energy/sector-framework.

IEA (International Energy Agency). 2020. "Global Investments in Oil and Gas Upstream in Nominal Terms and Percentage Change from Previous Year, 2010–2020." IEA, Paris, May. https://www.iea.org/data-and-statistics/charts/global-investments-in-oil-and-gas-upstream-in-nominal-terms-and-percentage-change-from-previous-year-2010-2020.

IFC (International Finance Corporation). 2017. "IFC's Definitions and Metrics for Climate-Related Activities." IFC Climate Business Department, Washington, DC. https://www.ifc.org/wps/wcm/connect/8ebdc507-a9f1-4b00-9468-7b4465806ecd/IFC+Climate+Definitions+v3.1+.pdf?MOD=AJPERES&CVID=lQuLLhw.

IFC (International Finance Corporation). 2019. "AIMM Sector Framework Brief." Oil & Gas, March, IFC, Washington, DC. https://www.ifc.org/wps/wcm/connect/98183cc6-fbf0-4cad-93cf-18c8ae2c7f66/AIMM-Oil-and-Gas-Consultation.pdf?MOD=AJPERES&CVID=mHZMjVh.

IsDB (Islamic Development Bank). 2018. "Energy Sector Policy: Sustainable Energy for Empowerment and Prosperity." IsDB, Jeddah, Saudi Arabia. https://www.isdb.org/sites/default/files/media/documents/2019-04/IsDB_Energy%20Sector%20Policy.pdf.

Loan Market Association. 2018. "Green Loan Principles: Supporting Environmentally Sustainable Economic Activity." Loan Market Association, London. https://www.lma.eu.com/application/files/9115/4452/5458/741_LM_Green_Loan_Principles_Booklet_V8.pdf.

Loan Market Association. 2021. "Powering Ahead: Shell's Innovative Approach to LIBOR Transition and Sustainability." Loan Market Association, London. https://www.lma.eu.com/application/files/3916/0519/2338/LMA_Interview_Shell_V01.pdf.

OGCI (Oil and Gas Climate Initiative). 2020. "Delivering on a Low Carbon Future: A Progress Report from the Oil and Gas Climate Initiative." OGCI, December. https://www.ogci.com/ogci-annual-reports/.

Preqin. 2020a. "2020 Preqin Global Natural Resources Report." Prequin, London.

Preqin. 2020b. "2020 Preqin Global Private Equity & Venture Capital Report." Prequin, London.

Preqin. 2020c. "Glossary of Terms." Prequin, London. https://vdocument.in/preqin-pro-glossary-of-terms-contents-private-equity-venture-capital-private-debt.html.

Rainforest Action Network. 2021. "Banking on Climate Chaos: Fossil Fuel Finance Report 2021." Rainforest Action Network, San Francisco. https://reclaimfinance.org/site/wp-content/uploads/2021/03/BOCC_2021_vF.pdf.

Takatsuki, Yo, and Julien Foll. 2019. "Financing Brown to Green: Guidelines for Transition Bonds." AXA Investment Managers Responsible Investment, June 10. https://realassets.axa-im.com/content/-/asset_publisher/x7LvZDsY05WX/content/financing-brown-to-green-guidelines-for-transition-bonds/23818#Management.

US Treasury. 2021. "Guidance on Fossil Fuel Energy at the Multilateral Development Banks." US Treasury, Washington, DC. https://home.treasury.gov/system/files/136/Fossil-Fuel-Energy-Guidance-for-the-Multilateral-Development-Banks.pdf.

World Bank. 2015a. "Common Principles for Climate Mitigation Finance Tracking (2nd Version)." World Bank, Washington, DC. http://www.worldbank.org/content/dam/Worldbank/document/Climate/common-principles-for-climate-mitigation-finance-tracking.pdf.

World Bank. 2015b. "What Are Green Bonds?" World Bank, Washington, DC. https://openknowledge.worldbank.org/handle/10986/22791.

Yamaguchi, Yuzo, and Rehan Ahmad. 2020. "Investors Applaud China's Plan to Ban Clean Coal from Green Bond Financing." S&P Global Market Intelligence, September 9. https://www.spglobal.com/marketintelligence/en/news-insights/latest-news-headlines/investors-applaud-china-s-plan-to-ban-clean-coal-from-green-bond-financing-60257794.

3 Financial Modeling of Gas Flaring and Methane Reduction Projects

This chapter provides a template for policy makers to evaluate the indicative financial attractiveness of flaring and methane reduction (FMR) projects by modeling net present values (NPVs) and internal rates of return (IRRs) of six FMR solutions: (1) gas-to-power, with power sold to the grid or other third-party off-takers; (2) gas-to-power, with power sold to the oil field operator for on-site use; (3) gas delivery to an existing pipeline network; (4) gas delivery to an existing gas processing plant; (5) compressed natural gas (CNG); and (6) small-scale liquefied natural gas (LNG). All financial modeling assumptions used in this chapter, although based on project experience, are indicative and do not reflect the wide variety of technological, contractual, regulatory, and end-market variables affecting FMR projects.

HIGHLIGHTS

The proposed models are intended as a high-level planning and assessment tool for policy makers to evaluate the potential to involve independent developers and private investors in FMR solutions. For the actual financing and implementation of specific FMR projects, developers and investors will likely want to use more granular and tailored assumptions.

The analysis points to a potentially attractive financial opportunity for independent developers to invest in FMR projects tackling flares of 5 million standard cubic feet per day (mmscf/d) or, better, 10 mmscf/d (figures 3.1 and 3.2). Indicative assumptions used in the financial models suggest the following:

- At 10 mmscf/d, all FMR solutions would produce positive NPVs and double-digit IRRs, ranging from 12 percent (gas delivery to gas processing plant) to 24 percent (small-scale LNG).

- At 5 mmscf/d flare sites, project IRRs—unlevered and pretax—range from a barely acceptable 7 percent[1] (gas delivery to gas processing plant) to an attractive 20 percent (small-scale LNG).

FIGURE 3.1

Summary of base case net present values

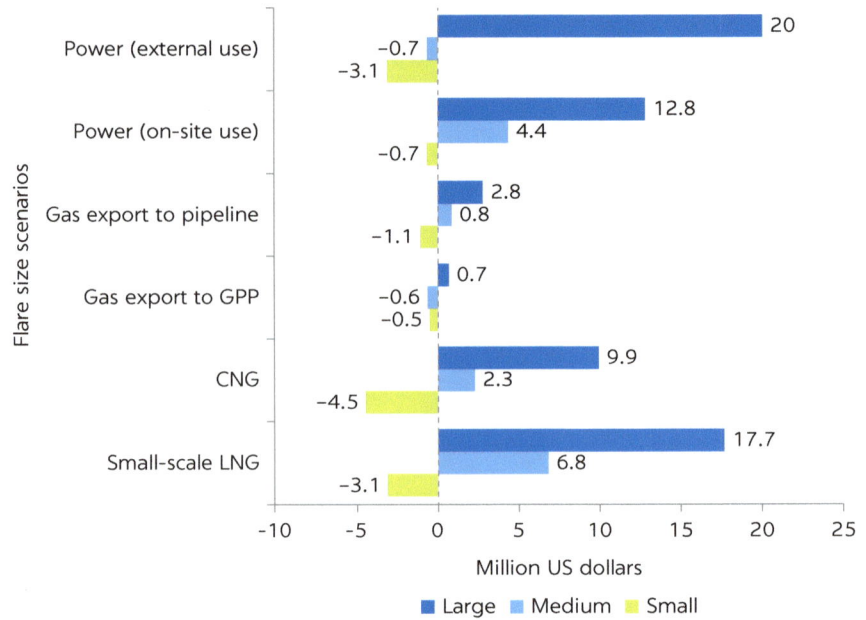

Source: World Bank.
Note: CNG = compressed natural gas; GPP = gas processing plant; LNG = liquefied natural gas.

FIGURE 3.2

Summary of base case internal rates of return

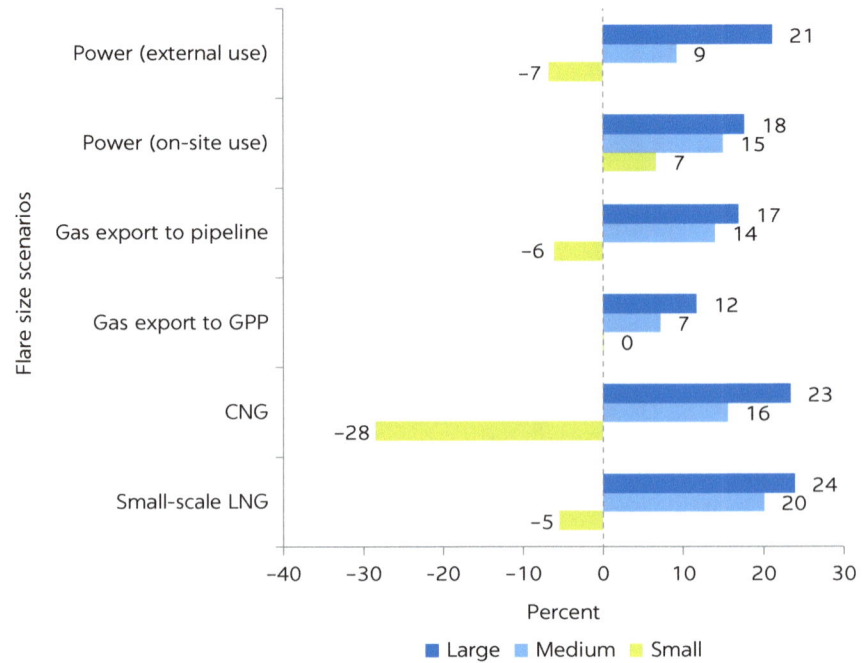

Source: World Bank.
Note: CNG = compressed natural gas; GPP = gas processing plant; LNG = liquefied natural gas.

- On a standalone basis, 1 mmscf/d flares do not offer attractive financial returns but can be clustered to reach an aggregate project size closer to 5–10 mmscf/d.

- FMR projects at 5–10 mmscf/d flare sites (unique flares or clusters) involve a capital investment estimated in the range of US$7 million to US$59 million, depending on the FMR solution adopted and according to the model's assumptions.

It must be emphasized that the financial models proposed in this chapter do not reflect each and every factor that may affect FMR returns. Such factors include implementation delays and cost overruns; complex negotiations with the oil producer and the off-taker; the competitive situation in local power or gas offtake markets; the volatility of end-product prices (power, gas, CNG, LNG) over the project period and availability of hedging tools; complications arising from the location of flare sites and the distance from end-product markets; and any local regulations creating additional compliance costs.

METHODOLOGY AND GENERAL ASSUMPTIONS

Base case assumptions are derived from publicly available information, complemented by interviews. Because of the varying conditions of each region and circumstances in which potential FMR projects occur, base case assumptions can only be indicative, and the same applies as a result to NPVs and IRRs. Appendix B includes a full print-out of the financial model. This section discusses general assumptions applicable to all solutions modeled. Additional solution-specific assumptions are discussed in the following sections.

Sensitivity analyses are provided to identify the parameters that have the greatest impact on project returns. For each FMR solution presented in this chapter, sensitivity of NPV and IRR to a range of variables is analyzed. In addition to highlighting financial risks to prospective FMR developers, the analysis also helps validate the robustness of conclusions emerging from the financial modeling exercise. In several FMR solutions, for instance, base case assumptions point to negative returns for projects taking place at small flare sites (1 mmscf/d).

The analysis in this chapter assumes that FMR projects are developed by a third-party project developer (see chapter 2), not the operator of the oil production facility, and that the developer aims to maximize the project's unlevered, pretax net NPV and IRR. It is assumed that developers finance FMR projects entirely through equity. This practice is often, but not always, the case because many FMR projects do not meet contractual project finance criteria—for instance, they may not have guaranteed feedstock supply agreements in place. In addition, tight development lines may limit the window of opportunity to contractually arrange a debt package. In some cases, however, depending on the project cashflow profile and available sources of debt in the project country, it is possible that FMR projects can be levered.

The financial model assumes zero inflation and no impact of exchange rate changes. NPV calculations are based on a 10 percent real discount rate, indicated by several industry participants as the floor equity return required considering FMR project-specific and macroeconomic risks. General assumptions are summarized in table 3.1 at the end of this section.

Because corporate tax regimes vary widely across countries, the base case simulations in this chapter are on a pretax basis. To show the impact of different tax rates, the sensitivity analysis for each of the modeled FMR solutions shows IRRs assuming a 10 percent and 20 percent corporate tax rate. The tax calculations assume straight-line depreciation of capital costs over five years, expensing of operating costs, and unlimited loss carry-forward, resulting in very limited impact of taxes on the financial returns of FMR projects with a base case lifetime of only seven years.[2]

Three associated gas volume scenarios—1 mmscf/d, 5 mmscf/d, and 10 mmscf/d—have been modeled. The three scenarios are referred to as "small," "medium," and "large" flare, respectively. In 2020, there were globally more than 2,611 upstream flare sites above 1 mmscf/d, representing 89 percent of the total flared gas volume and 36 percent of the total number of flare sites (figure 3.3).[3] For these flare sizes, oil producers have limited incentives to monetize the associated gas; FMR developers, by contrast, are likely to find the exercise worthwhile (see chapter 2).

As shown by the following analyses, flare and venting sites of less than 1 mmscf/d are generally very challenging to monetize. FMR developers will obviate this issue by clustering associated gas from small flares, leading to financial returns more akin to those witnessed for individual medium and large flares (figure 3.4). However, although still very low in absolute terms compared to typical oil development activities, the profitability of FMR investments significantly improves at the upper end of the flare size range considered in these evaluations.

FIGURE 3.3

Distribution of flared gas, by flare size, 2020

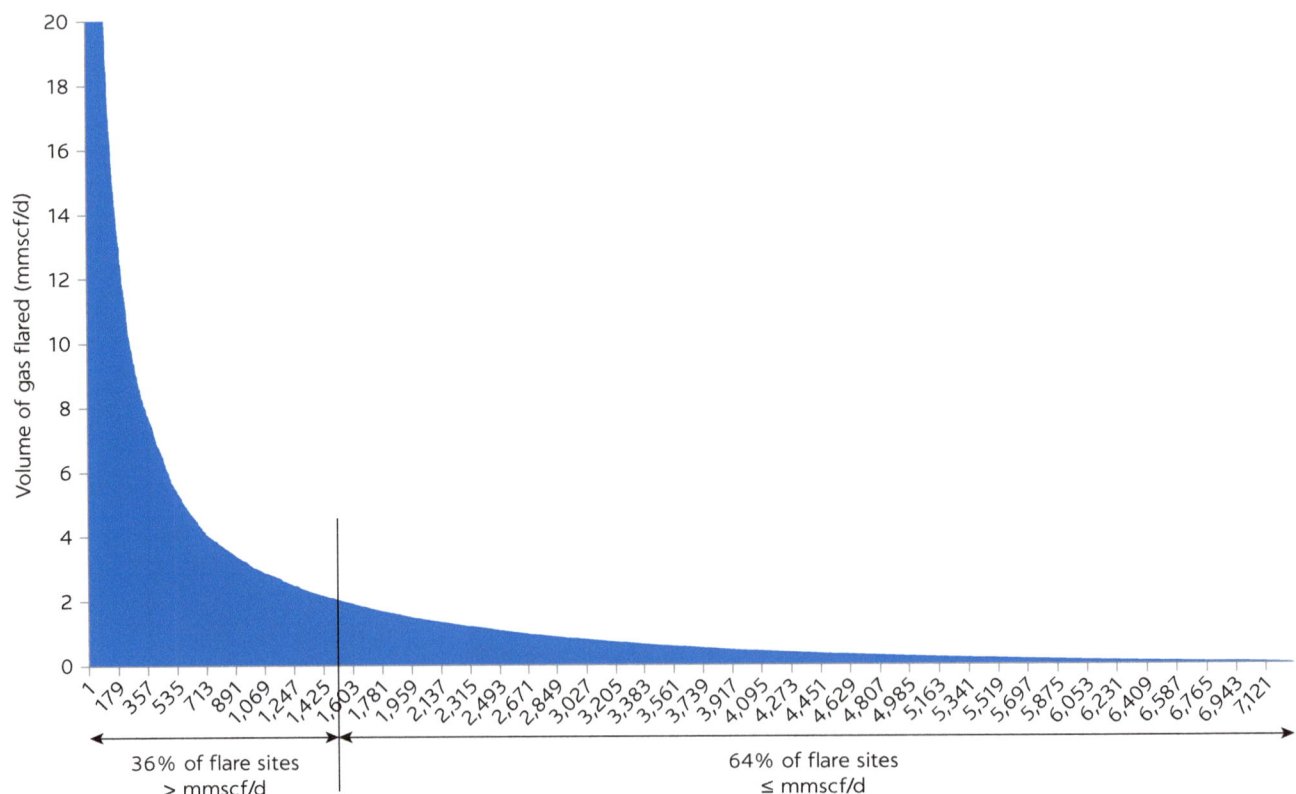

Source: Global Gas Flaring Reduction Partnership 2020 flare database (www.worldbank.org/ggfr).
Note: mmscf/d = million standard cubic feet per day.

FIGURE 3.4

Technical modalities of flare clustering

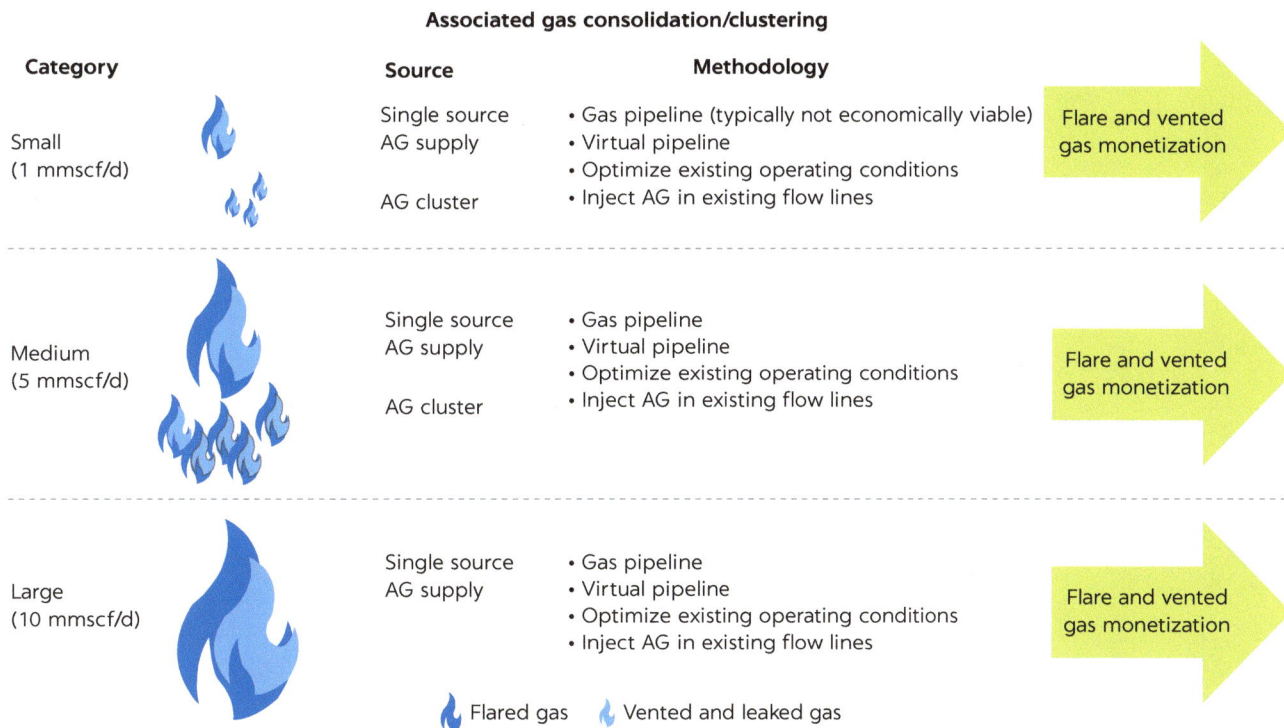

Source: World Bank.
Note: AG = associated gas; mmscf/d = million standard cubic feet per day.

It is further assumed that available gas volumes decline at 5 percent per annum—a rate typical of small, mature oil fields—and that 85 percent of this flared gas volume is routine flaring suitable for commercial use. Associated gas is composed mainly of methane but will often include water and heavier hydrocarbons (for example, ethane, propane, butane, and pentane). Depending on the type of gas use employed and the associated gas composition, the associated gas may require pretreatment (drying and removal of the heavier hydrocarbons as liquefied petroleum gas [LPG] and possibly carbon dioxide [CO_2] and hydrogen sulfide [H_2S]) before further processing. This pretreatment reduces the volume of gas available for further use or monetization and increases the cost and complexity of the gas processing. LPG or condensates extracted from the gas can, however, be valuable if sold separately or blended into the crude oil stream. Associated gas may sometimes contain nonhydrocarbon gases (contaminants) such as CO_2, H_2S, and nitrogen (N_2). In the analysis it is assumed that the gas contains sufficiently low concentrations of any of these contaminants that no pretreatment is required; if it did contain material volumes of any of these contaminants, in particular H_2S or CO_2, then additional treatment would generally be required, which can significantly increase the capital costs.

FMR projects are assumed to have a base case life of seven years (reflecting the period over which the gas supply is reasonably predictable and reliable, under conservative assumptions). To avoid oversizing the project capacity relative to the average gas availability, project capital expenditures (capex) are sized in line with

TABLE 3.1 **General assumptions**

ASSUMPTIONS	SMALL FLARE (1 mmscf/d)	MEDIUM FLARE (5 mmscf/d)	LARGE FLARE (10 mmscf/d)
Flare size (mmscf/d)	1	5	10
Routine flare portion (%)		85	
Annual decline rate (%)		5	
Project lifetime (years)		7	
Inflation	No inflation	No inflation	No inflation
Foreign currency impact	Stable exchange rates	Stable exchange rates	Stable exchange rates
Real discount rate (%)		10	
Tax rate (%)		0	
Project financing	100% equity	100% equity	100% equity

Source: World Bank.
Note: mmscf/d = million standard cubic feet per day.

the expected associated gas volume available at the end of year 3. Considering the inevitable associated gas decline during the project period, developers typically prefer to undersize equipment compared to the gas amount available on day 1, minimizing the risk of large overcapacity in the final years of the project when the operator's forecast gas volume will be significantly lower. The associated gas volume at the end of year 3 roughly coincides with the average volume available over seven years, at the assumed constant decline rate of 5 percent per annum.

To fully amortize equipment whose useful life is generally greater than seven years, FMR developers often opt for modular and movable technology that can be redeployed once the original flare is close to depletion. Collecting gas from a cluster of flares can be critical in making FMR projects financially attractive, especially in regions with multiple unstable flares, because it moderates short-term variations in the volume of gas supply, offers some economies of scale, and may extend a project's lifetime. Niche FMR solutions not covered in this chapter may be financially attractive in shorter time frames. Crusoe Energy Systems, for instance, uses associated gas to power data processing and Bitcoin mining data centers, using modular generation units and computing equipment and keeping the Bitcoins as reward or charging fees for power use. At April 2021 Bitcoin prices, Crusoe's solution generates attractive equity returns for flares as small as 0.3 mmscf/d of rich gas (see chapter 4 for details).

The IRR and NPV calculations presented in the following sections include sensitivities to project life, with an upside case of 10 years and a downside case of 5 years. Table 3.1 summarizes the key assumptions underlying the models presented in this chapter.

GAS-TO-POWER FOR THIRD-PARTY USE

Overview and assumptions

In this solution, an FMR developer installs and operates power generation equipment fueled with associated gas near the flare site and sells power to an existing local grid or other off-taker (figure 3.5). The model also assumes that the developer builds the necessary transmission lines, and pays for the associated

FIGURE 3.5

Schematics of gas-to-power (third-party use) solution

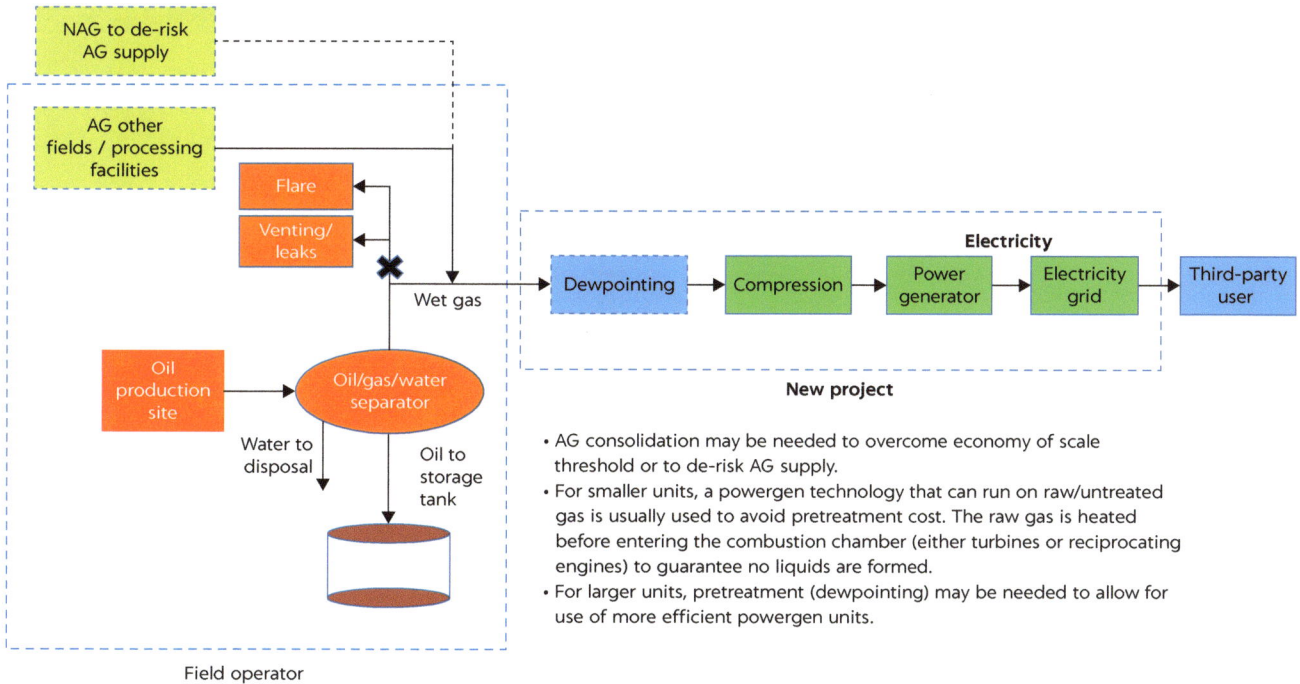

Source: World Bank.
Note: Nonassociated gas (NAG) is not essential in many projects. In some projects, however, NAG serves as a backup supply of gas, de-risking a project. AG = associated gas; powergen = power generation.

gas used, creating a small economic incentive for the oil field operator. Table 3.2 lists the assumptions specific to this scenario. More detailed technical assumptions on power generation efficiency, capex, and operating expenditures (opex) can be found in appendix B. The following paragraphs discuss the most important assumptions.

Power generation projects are capex-intensive but present significant economies of scale as the capacity installed increases. Small-scale power generation typically uses simple cycle reciprocating engines that, although of relatively low efficiency (in the range of 30–35 percent), are reliable and easy to maintain. Alternatively, in some cases turbines that can also run on raw gas have been used. Certain small reciprocating engines and turbines can run on unprocessed gas, but there is a tipping point at which, in exchange for added capex and opex to accommodate for dew point adjustment plants, one can run far more efficient prime movers (> 40 percent range in simple cycle). Based on the assumptions laid out in table 3.2, small, medium, and large flares will respectively enable 3, 17, and 38 megawatts of power generation capacity. Based on projects reviewed by the study team, the total development cost of power generation capex per megawatt can decrease from some US$2.5 million in the small flare scenario to US$1.5 million in the large flare scenario. Transmission capex (the cost of power lines) is trivial in comparison—in all three flare volume scenarios it represents no more than 5 percent of total capex, even for the longer distances assumed for the larger power plants.

The model's base case assumes that power is sold into the local electricity grid at prices in line with average global wholesale prices.[5] It should be noted that

TABLE 3.2 **Gas-to-power base case assumptions, third-party use**

MODEL VARIABLE	SMALL FLARE (1 mmscf/d)	MEDIUM FLARE (5 mmscf/d)	LARGE FLARE (10 mmscf/d)
Dew point adjustment	No (only shrinkage due to compression)	Yes (minimum dew point process to extract heavier hydrocarbons)	Yes (minimum dew point process to extract heavier hydrocarbons)
Dry gas after liquids/water removal (%)	98	93	93
Parasitic gas usage[a] (%)	3.0	2.5	2.0
Efficiency (Btu/kWh)	11,000	8,900	7,900
Gas energy content (Btu/scf)	1,050	996	996
Fuel gas needed for 1 MW of power generation capacity (mmscf/d)	0.25	0.20	0.18
Generation capacity installed (MW)	2.9	16.7	37.6
Powergen and gas handling capex per MW (US$)[4]	2,500,000	1,800,000	1,500,000
Powergen and gas handling capex (US$)	7,250,000	30,060,000	56,400,000
Power generation opex (% of capex)	11	12	10
Power plant availability (% of year)	94	94	94
Distance to grid/off-taker (km)	5	10	30
Transmission capex per km of overhead lines (US$)	60,000	100,000	100,000
Transmission capex (US$)	300,000	1,000,000	3,000,000
Transmission opex (% of capex)	2	2	2
Total capex (US$)	7,550,000	31,060,000	59,400,000
Wholesale electricity price (US$/kWh)	0.09	0.08	0.07
AG price paid to operator (US$/mscf)	0.25	0.25	0.25

Source: World Bank, based on data provided by GGFR.
Note: AG = associated gas; Btu = British thermal unit; capex = capital expenditures; kWh = kilowatt-hour; mmscf/d = million standard cubic feet per day; mscf = thousand standard cubic feet; MW = megawatt; opex = operating expenditures; powergen = power generation; scf = standard cubic feet.
a. Gas needed to run the compression and power generation equipment.

electricity prices vary greatly from country to country. FMR projects may also sell electricity to specific off-takers (such as individual industrial buyers) rather than to the local grid. Where grid supply is not available or its reliability is an issue, this practice may offer premium pricing opportunities.

Financial returns and sensitivities

Under base case assumptions, gas-to-power (third-party use) investments are financially attractive only for large flare sites (figure 3.6). The pretax, unlevered IRR of such investments is 9 percent in medium flares and 21 percent in large flares (figure 3.7). In the medium flare scenario, the IRR is below the 10 percent cost of equity assumed in this analysis, resulting in a slightly negative NPV.

Both IRR and NPV are substantially negative in the small flare scenario. Adjusting individual input variables favorably—increasing project life or electricity prices and decreasing capex, distance to grid, or associated gas price by the magnitudes shown in figure 3.8—does not result in positive IRRs. This suggests that gas-to-power (third-party use) investments are attractive in small flare sites only when a combination of variables is simultaneously favorable—for

FIGURE 3.6

Gas-to-power (third party): Base case net present value sensitivity

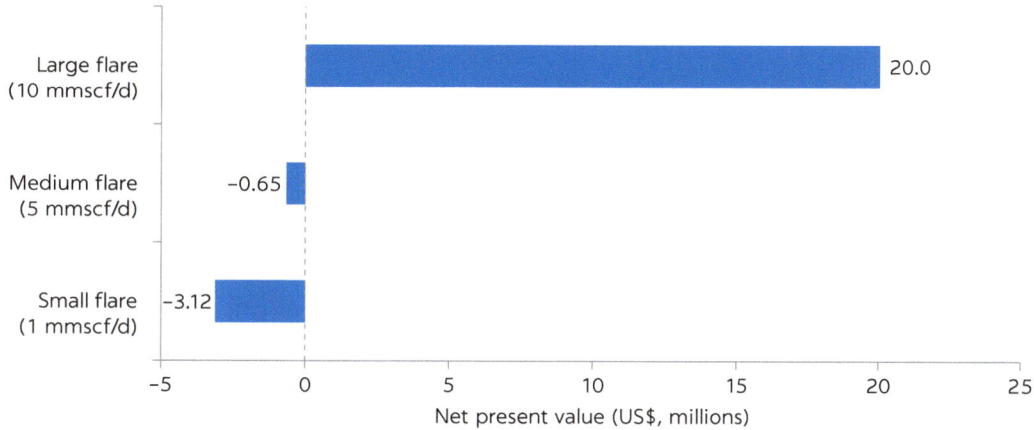

Large flare (10 mmscf/d): 20.0
Medium flare (5 mmscf/d): −0.65
Small flare (1 mmscf/d): −3.12

Net present value (US$, millions)

Source: World Bank.
Note: mmscf/d = million standard cubic feet per day.

FIGURE 3.7

Gas-to-power (third party): Base case internal rate of return sensitivity

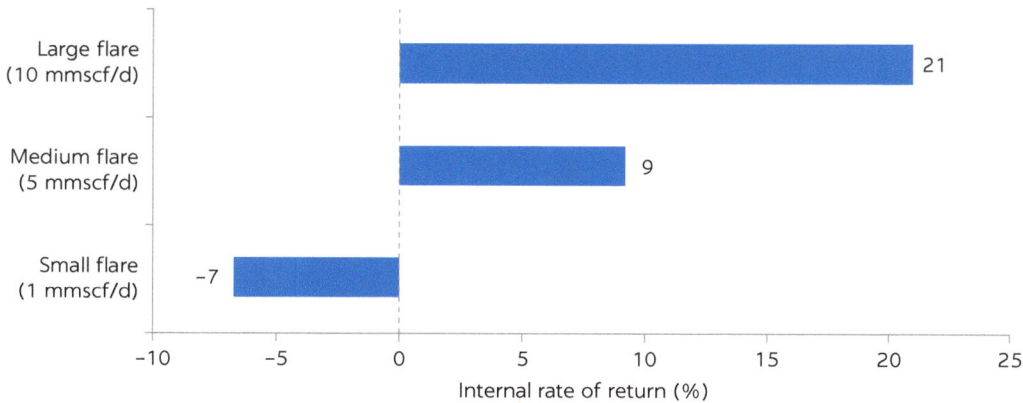

Large flare (10 mmscf/d): 21
Medium flare (5 mmscf/d): 9
Small flare (1 mmscf/d): −7

Internal rate of return (%)

Source: World Bank.
Note: mmscf/d = million standard cubic feet per day.

instance, when project life can be extended (perhaps by including gas from nearby flares when the volume of the original one has declined substantially) and electricity prices are very attractive.

Changes in project life, power generation capex, and power prices have the most significant impact on returns. Focusing on medium and large flares, the following impacts can be observed: increasing project life to 10 years adds 4–5 percentage points to the base case IRR; a 10 percent reduction in power generation capex adds 6 percentage points to the base case IRR; and a US$0.01 increase in power prices (to US$0.09 per kilowatt-hour [kWh]) adds 6–7 percentage points to the base case IRR. Similar results are observed when these variables are flexed in the opposite direction. By contrast, transmission capex and opex and associated gas price represent only a small fraction of project cashflows; as a result, a project that is 50 percent closer to the grid than in the base case or pays 50 percent less for associated gas is only marginally more attractive, with a 1-percentage-point increase in IRR.

FIGURE 3.8

Gas-to-power (third party): Internal rate of return under different scenarios

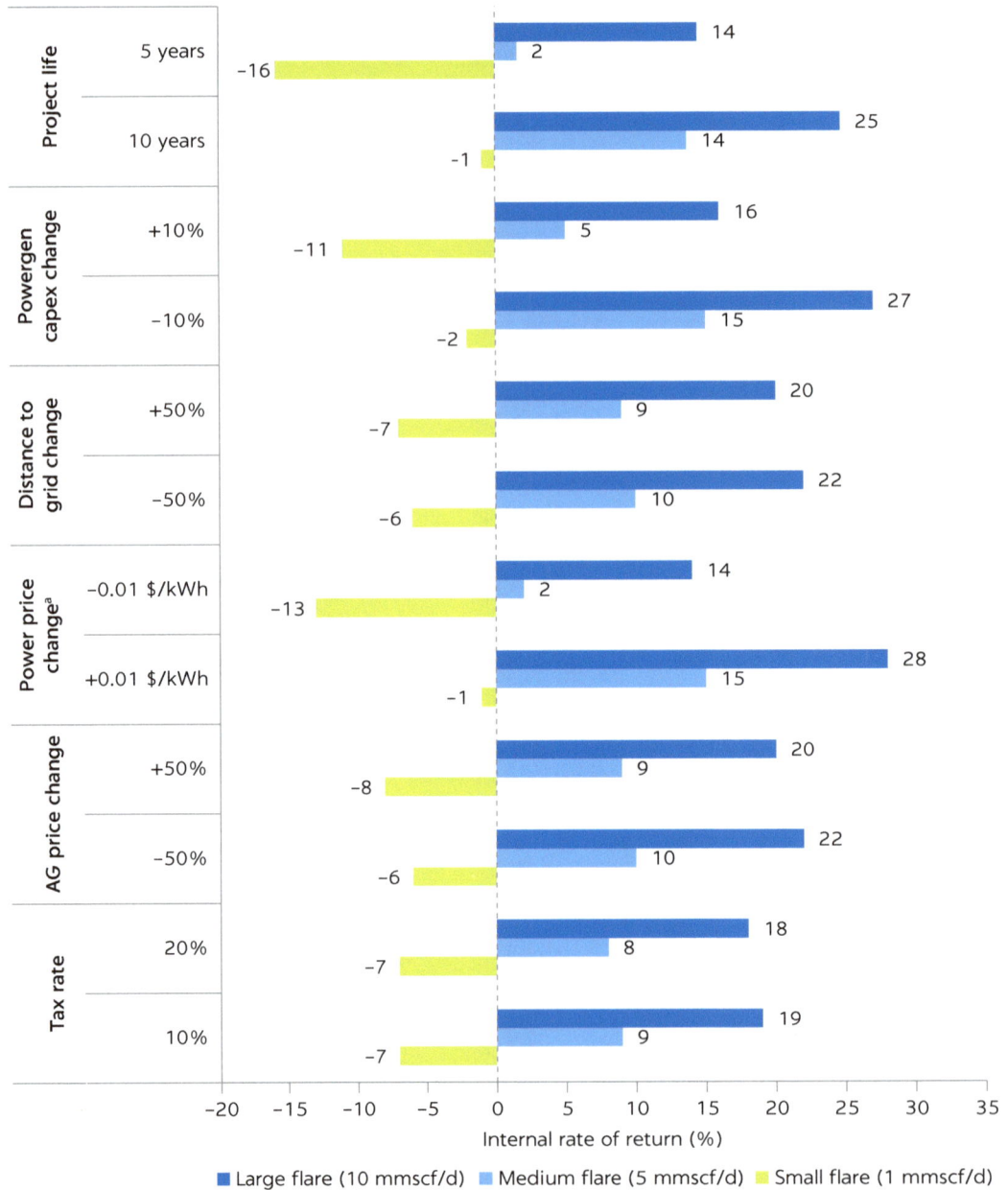

Source: World Bank.
Note: AG = associated gas; capex = capital expenditures; kWh = kilowatt-hour; mmscf/d = million standard cubic feet per day; powergen = power generation.
a. Bringing power prices to US$0.09 and 0.07/kWh respectively vs. base case of US$0.08/kWh.

GAS-TO-POWER FOR ON-SITE USE

Overview and assumptions

This solution differs from the gas-to-power for third-party use case discussed in the previous section by assuming that electricity generated from associated gas is used by the oil field operator for a tolling fee to displace electricity previously generated at the production facility using liquid fuel generators (figure 3.9).

FIGURE 3.9

Schematics of gas-to-power (on-site use) solution

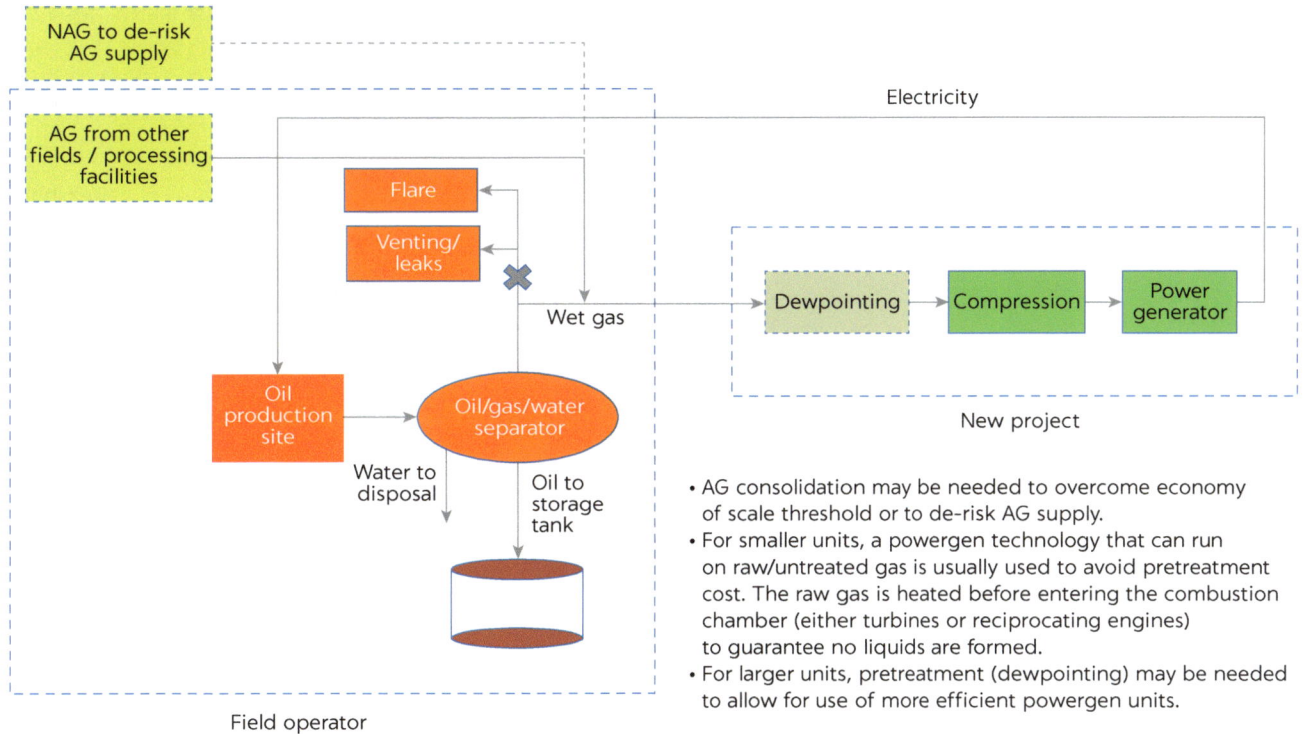

Field operator

- AG consolidation may be needed to overcome economy of scale threshold or to de-risk AG supply.
- For smaller units, a powergen technology that can run on raw/untreated gas is usually used to avoid pretreatment cost. The raw gas is heated before entering the combustion chamber (either turbines or reciprocating engines) to guarantee no liquids are formed.
- For larger units, pretreatment (dewpointing) may be needed to allow for use of more efficient powergen units.

Source: World Bank.
Note: Nonassociated gas (NAG) is not essential in many projects. In some projects, however, NAG serves as a backup supply of gas, de-risking a project. AG = associated gas; powergen = power generation.

This scenario is unlikely to apply when operators have already met all of their power demand with associated gas produced by the oil field(s). Because electricity from liquid-fueled generators costs typically about US$0.20/kWh, the electricity price charged by the FMR developer can be higher than in the gas-to-power for third-party use scenario.[6] In that scenario, competition from low-cost alternative electricity providers typically limits potential realized prices.

All assumptions in this scenario are the same as in the third-party use scenario, with the following exceptions:

- Power prices are assumed to be US$0.10/kWh in the small flare case, decreasing to US$0.085/kWh and US$0.07/kWh in the medium and large flare cases, respectively. The decrease reflects the higher bargaining power of the oil field operator for larger volumes of electricity purchase.[7]

- Associated gas is assumed to be supplied at no cost by the operator to the FMR developer. The tolling fee paid by the operator to the FMR developer is assumed to take account of the zero cost of the associated gas.

- The negligible capital and operating costs of the very short transmission lines are assumed to be zero for simplicity.

Financial returns and sensitivities

Under base case assumptions, gas-to-power (on-site use) investments produce attractive IRRs and NPVs in medium and large flares. In small flares, the IRR is positive but below the assumed 10 percent cost of equity, resulting in a negative NPV (figure 3.10). Pretax, unlevered IRRs are 7 percent, 15 percent, and 18 percent in small, medium, and large flares, respectively (figure 3.11). The addition of leverage, if available at interest rates lower than the unlevered IRR, or favorable change in at least one of the core assumptions would be sufficient to increase the IRR to double digits in the small flare case.

Changes in project life, power generation capex, and power prices have the most significant impact on returns. An increase in project life to 10 years adds 3–4 percentage points to the base case IRR; a 10 percent reduction in power generation capex relative to the base case adds 5–6 percentage points; and a

FIGURE 3.10

Gas-to-power (on-site use): Base case net present value sensitivity

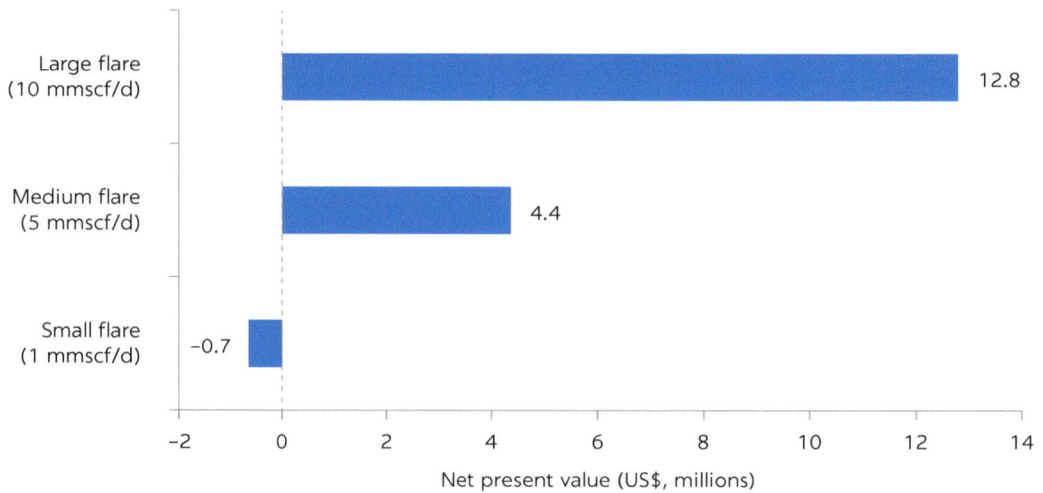

Source: World Bank.
Note: mmscf/d = million standard cubic feet per day.

FIGURE 3.11

Gas-to-power (on-site use): Base case internal rate of return sensitivity

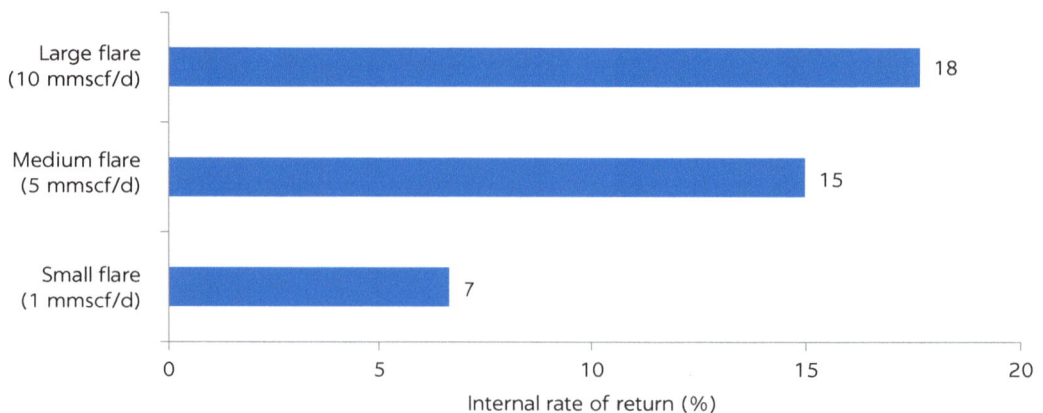

Source: World Bank.
Note: mmscf/d = million standard cubic feet per day.

US$0.01/kWh increase in power prices adds 4–7 percentage points to the IRR. The same movements occur symmetrically when flexing the capex and power price assumptions to the downside by the same order of magnitude. Cutting project life to 5 years has a more pronounced negative impact on IRR—a drop of 7–8 percentage points—because of a smaller number of positive-cashflow years to offset the high negative cashflow (capex-driven) at project inception (figure 3.12).

FIGURE 3.12

Gas-to-power (third party): Internal rate of return under different scenarios

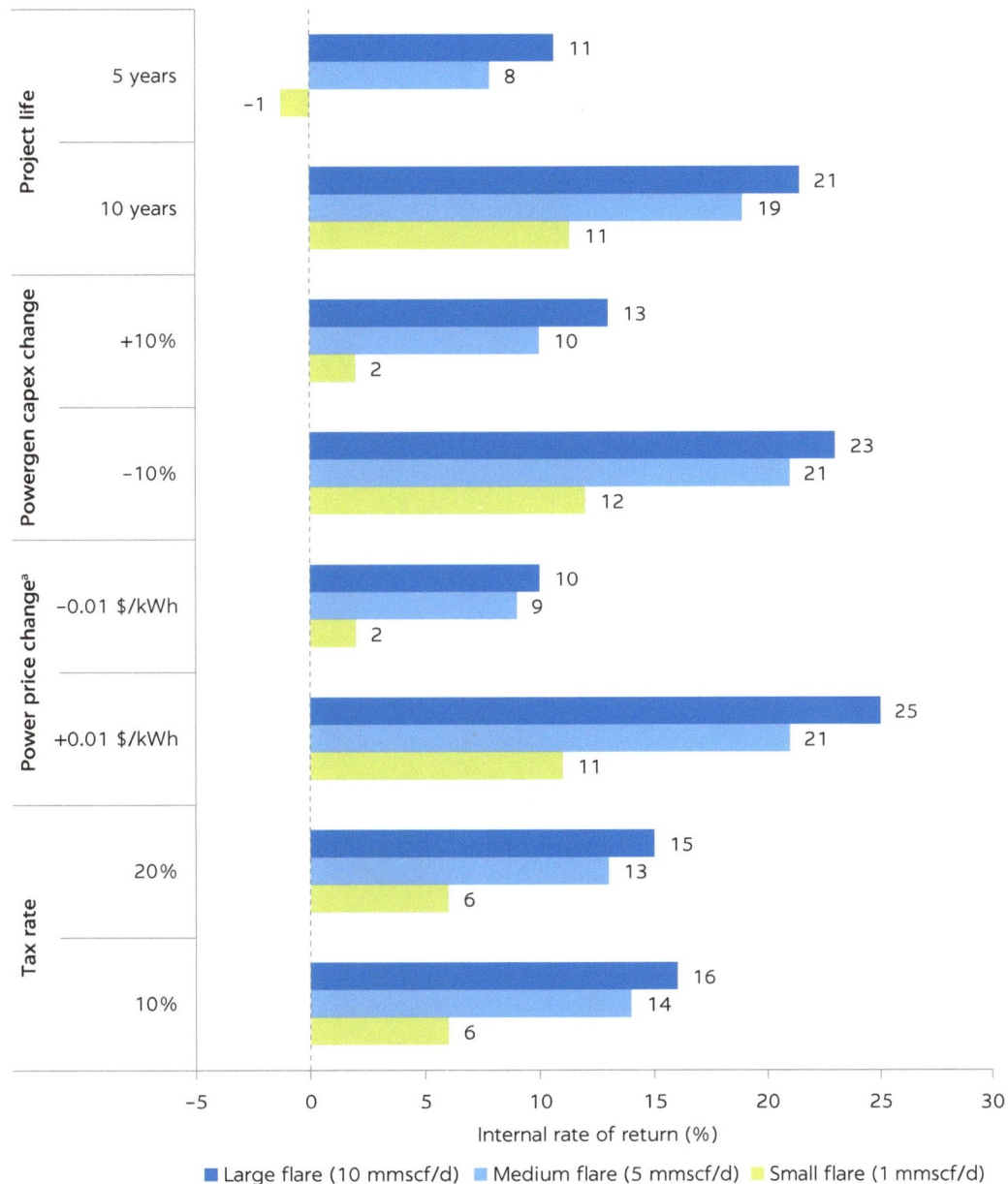

Internal rate of return (%)

■ Large flare (10 mmscf/d) ■ Medium flare (5 mmscf/d) ■ Small flare (1 mmscf/d)

Source: World Bank.
Note: capex = capital expenditures; kWh = kilowatt-hour; mmscf/d = million standard cubic feet per day; powergen = power generation.
a. Above or below base case prices of US$0.10/kWh, US$0.085/kWh, and US$0.07/kWh for small, medium, and large flares, respectively.

GAS DELIVERY TO EXISTING PIPELINE NETWORK

Overview and assumptions

In this solution, an FMR developer installs and operates gas pretreatment and compression equipment and a pipeline to supply and sell gas into an existing pipeline network (figure 3.13). Gas delivered to an existing gas pipeline infrastructure must meet the specifications of the infrastructure owner. This requirement means that the gas must be pretreated to pipeline specifications—for instance, gas must be dry and have a controlled heating value—by removing the heavier hydrocarbons and drying the gas before compression. The dry gas is then compressed from 40 pounds per square inch (psi), typical outlet of the separator or gas pretreatment, to > 300 psi for delivery by pipeline. The FMR developer (1) invests in the liquid removal, dew point adjustment, and compression equipment as well as the pipeline connecting the pretreatment and compression plant to the existing pipeline infrastructure; (2) is responsible for operations and maintenance; and (3) is remunerated via tolling fees paid by the oil field operator, which receives revenues from the sale of gas. Table 3.3 lists all assumptions specific to this solution. The following paragraphs discuss the most critical assumptions.

Capex for LPG removal, dew point adjustment, and compression benefit from significant economies of scale. The investment required to treat and compress each mmscf of associated gas in the large flare scenario is less than half on a US dollar per mmscf/d basis than that of a small flare scenario (US$1.6 million per mmscf/d vs. US$3.3 million per mmscf/d). As a result, although a large flare

FIGURE 3.13

Schematics of gas delivery to existing pipeline network

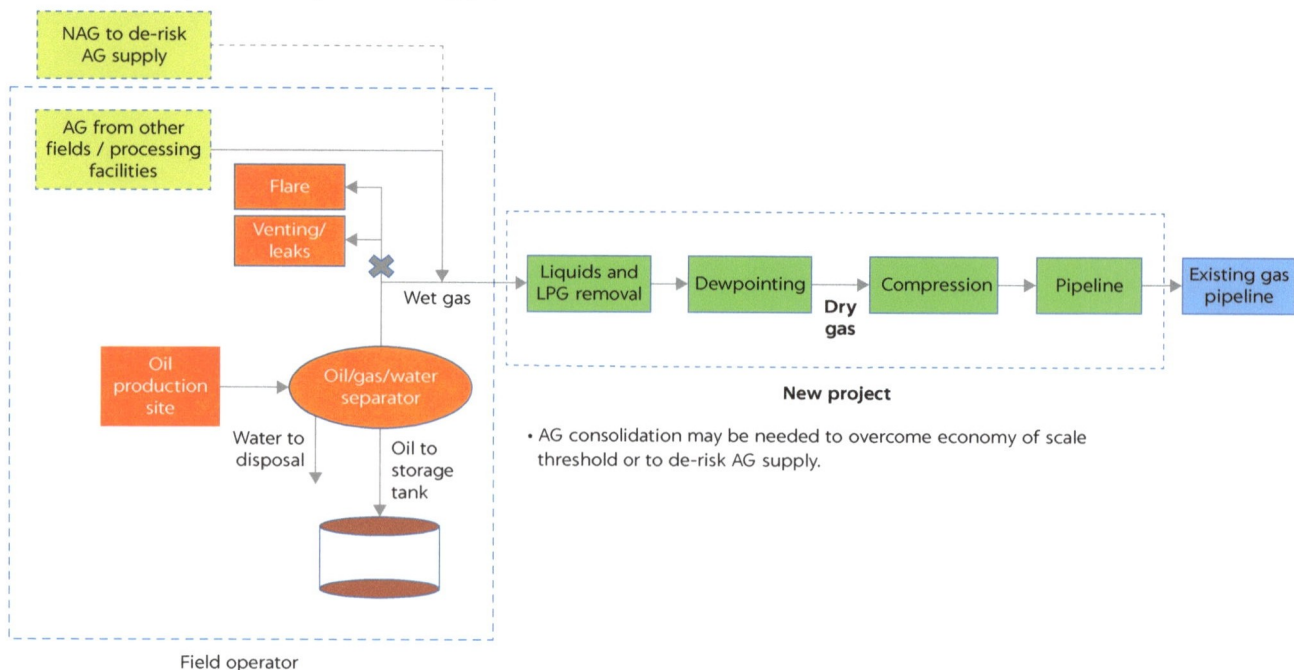

Source: World Bank.
Note: Nonassociated gas (NAG) is not essential in many projects. In some projects, however, NAG serves as a backup supply of gas, de-risking a project.
AG = associated gas; LPG = liquefied petroleum gas.

TABLE 3.3 **Assumptions regarding gas delivery to existing pipeline network**

ASSUMPTIONS	SMALL FLARE (1 mmscf/d)	MEDIUM FLARE (5 mmscf/d)	LARGE FLARE (10 mmscf/d)
Dry gas after liquids/water removal (%)	90	90	90
Parasitic gas usage (%)[a]	13	13	13
Capex for LPG removal, dew point adjustment, and compressor (US$/mmscf/d)[b]	3,280,000	1,780,000	1,580,000
Capex for LPG removal, dew point adjustment, and compressor (US$)[c]	2,296,000	5,874,000	10,428,000
Distance to existing pipeline network (km)	1	3	5
Capex for connection pipeline (US$/km)	300,000	450,000	600,000
Capex for connection pipeline (US$)	300,000	1,350,000	3,000,000
Total capex (US$)	2,596,000	7,224,000	13,428,000
Opex (as % of capex)	3.5	3.5	3.5
LPG extraction, dew point, and compressor availability (% of year)	98	98	98
Tolling fee (US$/mscf)	2.00	2.00	2.00
LPG sales price (US$/mscf)	3.00	3.00	3.00

Source: World Bank, based on data provided by Global Gas Flaring Reduction Partnership.
Note: capex = capital expenditures; km = kilometer; LPG = liquefied petroleum gas; mmscf/d = million standard cubic feet per day; mscf = thousand standard cubic feet; opex = operating expenditures.
a. Seven percent to power LPG extraction, 3 percent to run dew point adjustment, and 3 percent to run compressor.
b. Cost estimates to compress gas based on 2020 data provided by compressor manufacturers Ariel and Sertco, with an additional allowance for installation and civil works.
c. Based on gas treated and compressed per day in year 3 of the project, approximately equivalent to the average across the seven-year project period.

is 10 times as large as a small flare (10 mmscf/d vs. 1 mmscf/d), its treatment and compression capex is only five times as large.

Because of the high cost of building a connection pipeline, for a project to be economically viable the flare site should be near existing gas transport infrastructure. The base case scenario assumes distances of 1, 3, and 5 kilometers (km) for small, medium, and large flares, respectively. A larger volume of gas can justify longer delivery distances but is (partly) offset by the increased diameter and therefore unit cost of the pipes required. The model assumes that pipeline capex per kilometer for a large flare is twice that for a small flare (US$600,000/km vs. US$300,000/km).

The model assumes the pipeline is operated on a tolling fee basis (US$2.00 per thousand standard cubic feet [mscf]), eliminating the need to purchase and sell the gas. The tolling fee negotiated with the oil field operator reflects the investment made by the FMR developer for both gas treatment/compression and transport infrastructure; it is therefore assumed to be higher than tolling fees applicable only for the operation of transport infrastructure.

The ability to sell the LPG extracted before injecting gas into the pipeline can significantly improve the return profile of this FMR solution. The model's base case assumes no sale of LPG. The sensitivity analysis, however, includes a simulation of LPG sale. In this scenario, LPG separated from the wet gas would be spiked into the oil export line (if possible) and sold with the crude at an indicative price of US$3/mscf (equivalent to US$3 per million British thermal unit [mmBtu], assuming that 1 scf is approximately 1,000 Btu), which would reflect a discount of approximately 50 percent at an indicative oil price of US$40 per barrel (a conservative price compared to oil market prices as of May 2021). In practice, the FMR developer's ability to monetize LPGs and the split of LPG revenue will need to be negotiated with the oil field operator.

Financial returns and sensitivities

Under base case assumptions, the delivery of gas to an existing pipeline infrastructure is financially attractive only for medium and large flare sites. The pretax, unlevered IRR of such projects is 14 percent in medium flares and 17 percent in large flares (figure 3.14), and NPVs are positive. Both the IRR and NPV are substantially negative in the small flare scenario, reflecting the disproportionately high cost of gas treatment, compression, and transport infrastructure compared to the volumes of gas available. Flexing individual variables favorably—increasing project life or tolling fees, or decreasing capex or distance to existing infrastructure by the magnitudes shown in figure 3.15—does not result in a positive IRR or NPV in the small flare scenario. This finding suggests that this solution is attractive in small flare sites only when a confluence of positive factors occurs simultaneously.

In medium and large flares, extending project life to 10 years, decreasing treatment and compression capex by 10 percent, decreasing the distance to the existing pipeline by 50 percent, or increasing tolling fees by 10 percent would each add 3–5 percentage points to the IRR. Changing any one of these

FIGURE 3.14

Pipe-to-pipe: Base case net present value sensitivity

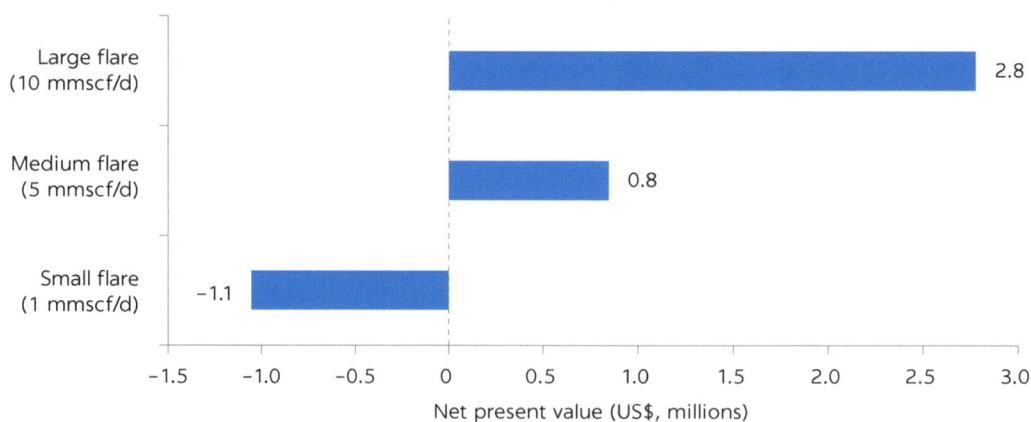

Source: World Bank.
Note: mmscf/d = million standard cubic feet per day.

FIGURE 3.15

Pipe-to-pipe: Base case internal rate of return sensitivity

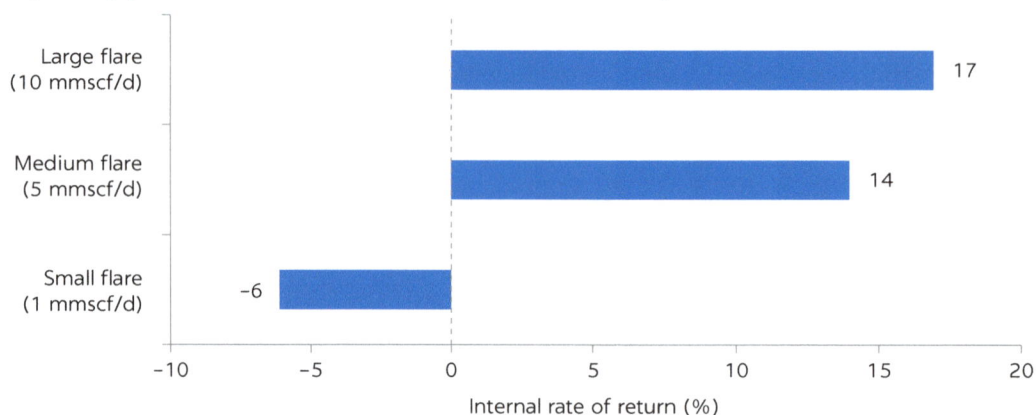

Source: World Bank.
Note: mmscf/d = million standard cubic feet per day.

assumptions would increase the IRR to 18–19 percent and 21–22 percent in the medium and large flares, respectively. Similar movements would happen symmetrically when changing the assumptions to the downside. As in previous scenarios, cutting project life to 5 years has a more pronounced negative impact on IRR, with a drop of 7–8 percentage points (figure 3.16).

The ability of the FMR developer to fully monetize LPGs at US$3/mscf would have a major positive impact on financial returns. The IRR would increase by 6–7 percentage points, to 20 percent and 24 percent, respectively, in the medium and large flare scenarios. The sale of LPGs would still be insufficient to bring the small flare IRR into positive territory—a testament to the difficulty of making this FMR solution work at a small scale.

FIGURE 3.16

Pipe-to-pipe: Internal rate of return under different scenarios

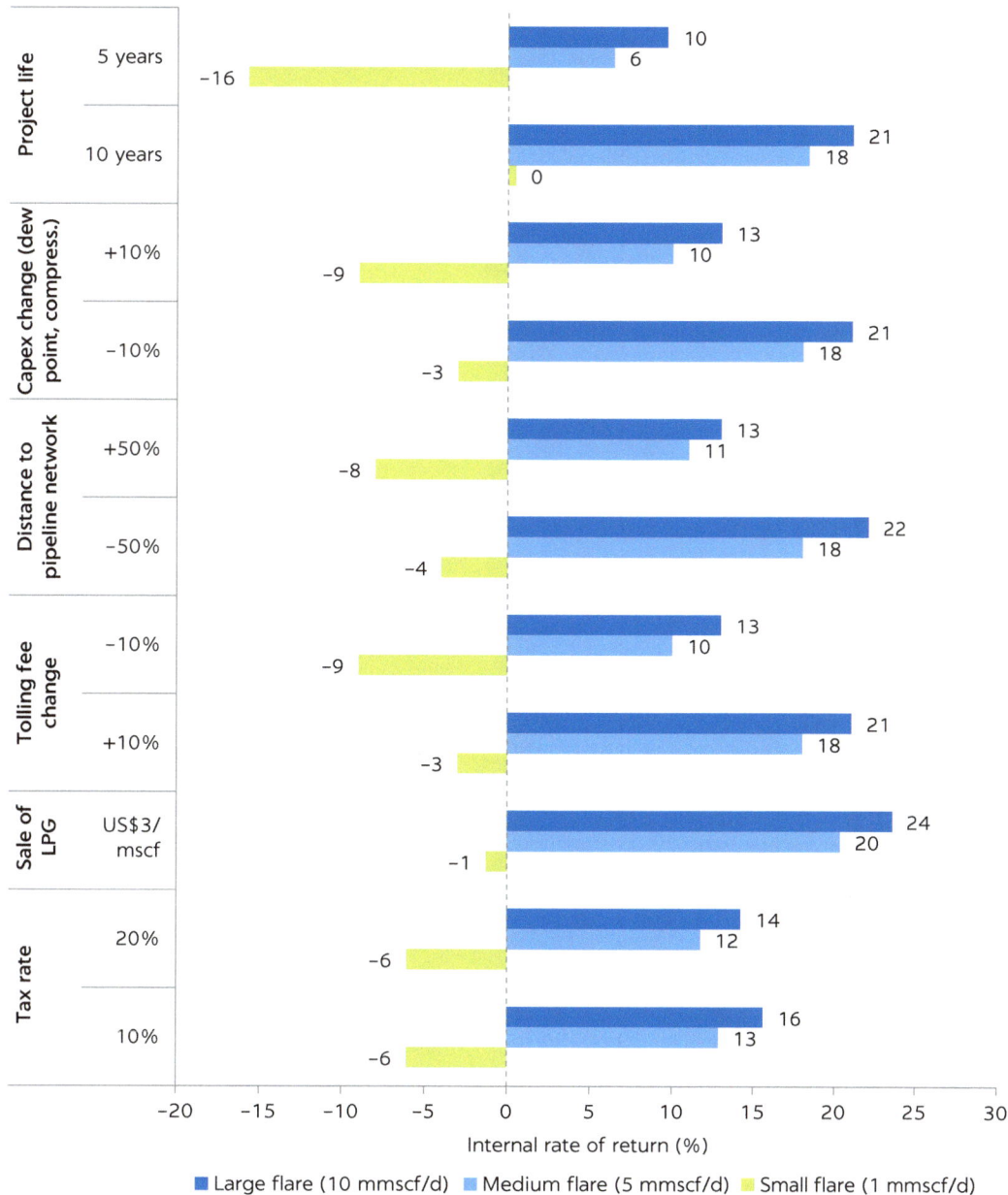

Source: World Bank.
Note: capex = capital expenditures; LPG = liquefied petroleum gas; mmscf = million standard cubic feet per day.

GAS DELIVERY TO EXISTING GAS PROCESSING PLANT

Overview and assumptions

In this solution, an FMR developer installs and operates compression equipment and a pipeline to deliver gas to an existing gas processing plant (GPP; figure 3.17). This scenario applies the same assumptions used in table 3.3, except as specified in table 3.4.

FIGURE 3.17

Schematics of gas delivery to existing gas processing plant

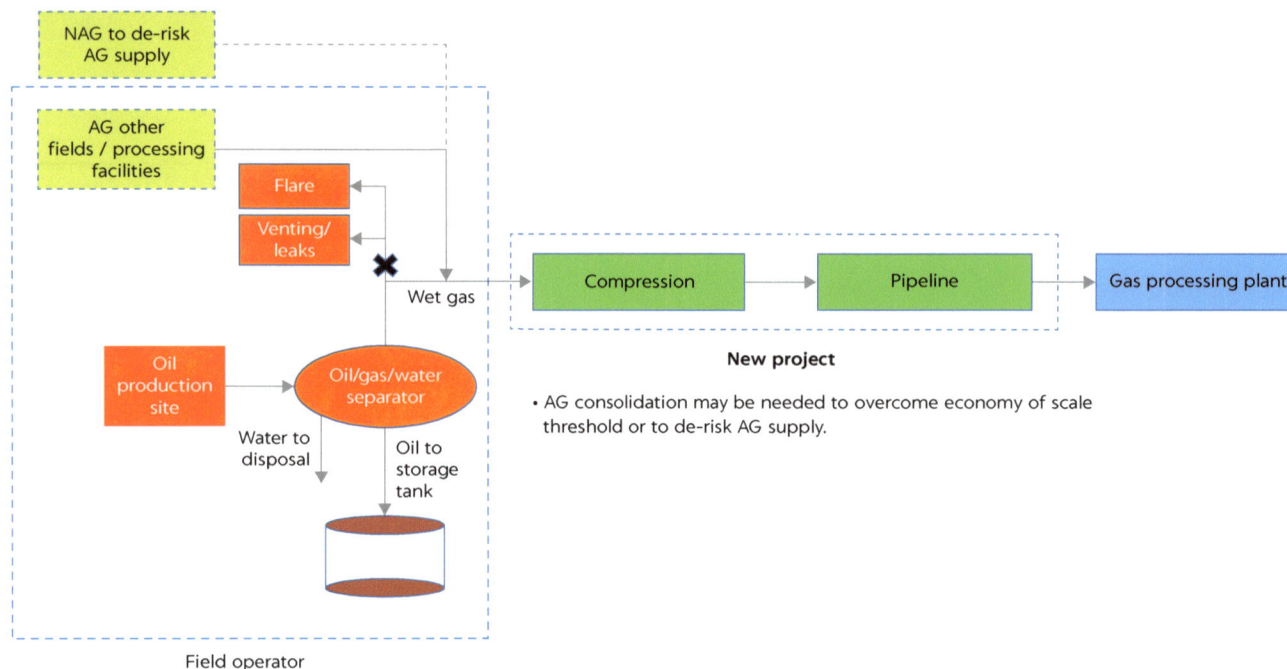

Source: World Bank.
Note: Nonassociated gas (NAG) is not essential in many projects. In some projects, however, NAG serves as a backup supply of gas, de-risking a project.
AG = associated gas; LPG = liquefied petroleum gas.

TABLE 3.4 **Assumptions regarding gas delivery to existing gas processing plant**

ASSUMPTIONS	SMALL FLARE (1 mmscf/d)	MEDIUM FLARE (5 mmscf/d)	LARGE FLARE (10 mmscf/d)
Capex for compressor (US$/mmscf/d)[a]	600,000	350,000	310,000
Capex for compressor (US$)[b]	480,000	1,295,000	2,263,000
Distance to existing pipeline network (km)	5	15	20
Capex for connection pipeline (US$)	1,500,000	6,750,000	12,000,000
Total capex (US$)	1,980,000	8,045,000	14,263,000
Tolling fee (US$/mscf)	1.50	1.50	1.50

Source: World Bank, based on data provided by GGFR.
Note: capex = capital expenditures; km = kilometer; mmscf/d = million standard cubic feet per day; mscf = thousand standard cubic feet.
a. Cost estimates to compress gas based on 2020 data provided by compressor manufacturers Ariel and Sertco, with an additional allowance for installation and civil works.
b. Based on gas compressed per day in year 3 of the project, approximately equivalent to the average across the seven-year project period.

This solution differs from the previously shown gas delivery to existing gas pipeline scenario in the following respects:

- Capex requirements at the flare site are lower, because gas delivered to an existing GPP can be delivered as wet, untreated gas provided it contains only minor levels of CO_2 or H_2S. The gas therefore requires no pretreatment to remove any liquids or contaminants before the compression. This allowance reduces capex requirements at the flare site to the purchase and installation of a compressor, an investment estimated to range between approximately US$0.5 million (small flare) and US$2.3 million (large flare).

- Capex requirements for the construction of a pipeline connecting the flare site to the GPP are higher, because a GPP is less likely to be in the proximity of a flare site and the pipeline length required is therefore assumed to be greater. The model assumes distances to a GPP of 5, 15, and 20 km for small, medium, and large flares, respectively (a larger volume of gas justifies longer distances). At the same cost per kilometer as in the previous solution, pipeline capex ranges from US$1.5 million (small flare) to US$12 million (large flare).

- Wet gas is assumed to be delivered for a tolling fee of US$1.50/mscf. This price is lower than that assumed for the delivery of clean, dry gas to existing gas infrastructure because the lack of processing of the raw gas reduces the project's capital costs, as previously noted.

- Because this scenario involves no extraction of liquids, there is no LPG to potentially monetize.

Financial returns and sensitivities

Under base case assumptions, the delivery of gas to an existing GPP is financially attractive only for flare sites above 5 mmscf/d (figure 3.18). The pretax, unlevered IRR of such investments is 7 percent in medium flares and 12 percent in large flares (figure 3.19). IRR is at breakeven and NPV negative in the small flare scenario, reflecting the disproportionately high investment in gas compression and transport infrastructure compared to the volumes of gas available.

Under the base case assumptions in this analysis, returns in the medium and small flare scenarios will likely be lower than in the case of delivery to a pipeline, which also requires gas treatment. Tolling fee negotiations will likely be very tight as a result, without the added bonus of LPG revenues. The 7 percent (medium flare) and 12 percent (large flare) IRRs for delivery to a GPP compare to 14 percent and 17 percent IRRs (respectively) in the delivery to pipeline solution.

Financial returns are most affected by the flare site's proximity to a GPP, given the weight of pipeline construction capex in the total project capex. In the case of a 50 percent reduction in distance to a GPP, IRRs in all scenarios including small flare would be extremely attractive (18 percent for a small flare, 30 percent and more for medium and large flares). In practice, as previously noted, it is highly unlikely that a GPP is situated in very close proximity to a flare site. Even if that were the case, a rational oil field operator would likely negotiate lower tolling fees than the assumed US$1.50/mscf, rather than generating a financial windfall for the FMR developer. Flexing one of the other variables shown in figure 3.20 results in an IRR change in the range of 3–5 percentage points.

FIGURE 3.18

FIGURE 3.18

Pipeline to gas processing plant: Base case net present value sensitivity

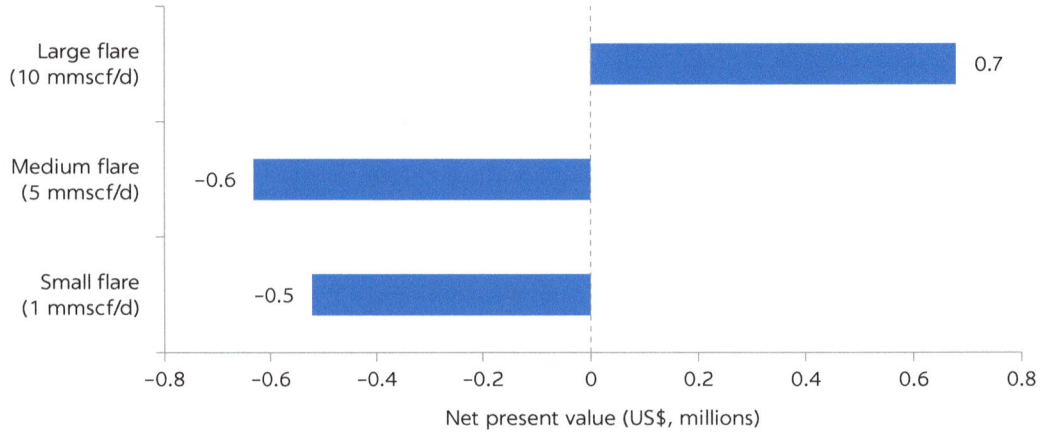

Source: World Bank.
Note: mmscf/d = million standard cubic feet per day.

FIGURE 3.19

Pipeline to gas processing plant: Base case internal rate of return sensitivity

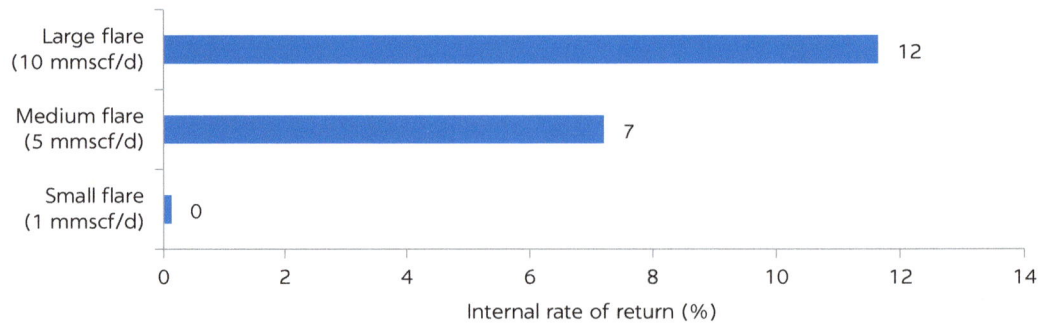

Source: World Bank.
Note: mmscf/d = million standard cubic feet per day.

COMPRESSED NATURAL GAS

Overview and assumptions

In this solution, an FMR developer invests in and operates equipment to pretreat and compress associated gas into compressed natural gas (CNG) and trucks to transport CNG to the end buyer, who pays market prices (figure 3.21). CNG projects comprise compressing the gas to ~3,000 psi to increase its energy density and thereby reduce the cost per unit of transporting the gas. The majority of CNG projects are virtual pipelines involving gas delivery by truck to a user, typically for power generation or heating (often displacing diesel) or—less likely—injection into existing gas infrastructure. It is also possible to use the CNG directly as an alternative fuel for cars, vans, and small trucks. These applications require the gas to be pretreated to pipeline specifications (including heavier hydrocarbon removal and controlled heating value) before compression. Compression is required to take the gas from approximately 40 psi (typical gas

FIGURE 3.20

Pipeline to gas processing plant: Internal rate of return under different scenarios

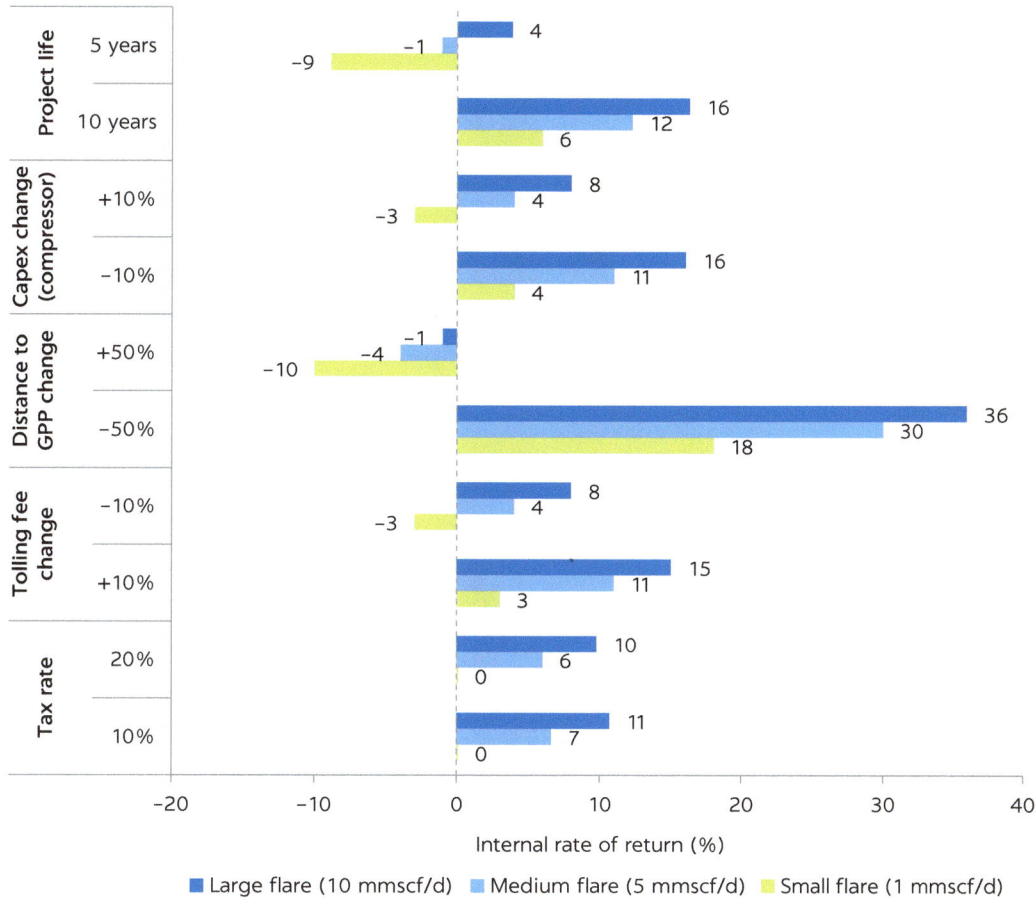

Source: World Bank.
Note: capex = capital expenditures; GPP = gas processing plant; mmscf/d = million standard cubic feet per day.

pressure at the outlet of the gas treatment plant) to a CNG container pressure of about 3,000 psi. This analysis assumes that the project is land-based and the CNG is transported by road in pressurized containers. It also assumes that equipment required for use of the CNG by the CNG buyer (for example, container blow-down) is purchased by the buyer, not the FMR developer. The project incurs operating costs for preprocessing, compression, and delivery of the CNG by truck to a maximum distance of 150 km and for the purchase of raw associated gas from the oil field operator. Revenues are derived from selling CNG at prevailing market prices. Table 3.5 lists all assumptions specific to this solution. The following paragraphs discuss the most critical assumptions.

Capex for LPG removal, dew point adjustment, compression, and trucks and CNG containers benefit from significant economies of scale. The investment required to treat, compress, and transport 1 mmscf/d of associated gas in the large flare scenario is less than half that of the small flare scenario (US$3.7 million vs. US$8.8 million). As a result, although a large flare for the purpose of this model is 10 times as large as a small flare (10 mmscf/d vs. 1 mmscf/d), the required capex is only 4 times as large.

The need to operate a fleet of trucks, in addition to gas treatment and compression equipment, results in high opex (13 percent of total capex per

FIGURE 3.21

Schematics of compressed natural gas solution

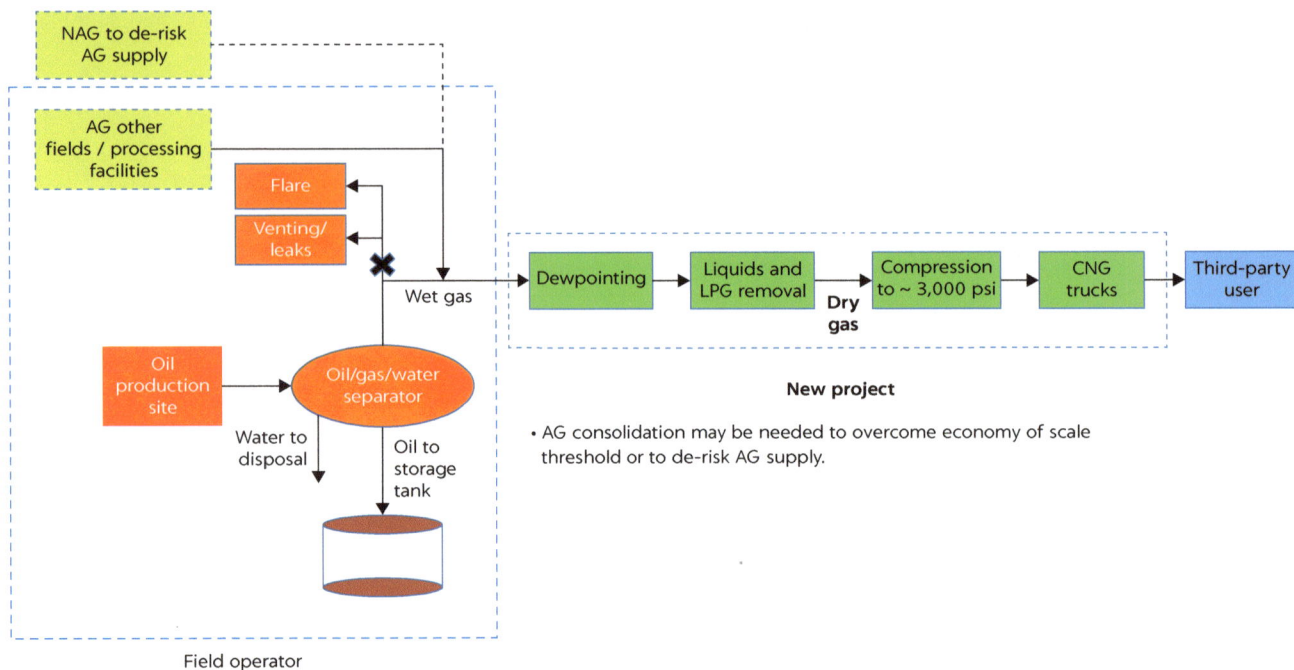

Source: World Bank.
Note: Nonassociated gas (NAG) is not essential in many projects. In some projects, however, NAG serves as a backup supply of gas, de-risking a project. AG = associated gas; CNG = compressed natural gas; LPG = liquefied petroleum gas; NAG = nonassociated gas; psi = pounds per square inch.

TABLE 3.5 **Assumptions regarding compressed natural gas scenario**

ASSUMPTIONS	SMALL FLARE (1 mmscf/d)	MEDIUM FLARE (5 mmscf/d)	LARGE FLARE (10 mmscf/d)
Dry gas after liquids/water removal (%)	90	90	90
Parasitic gas usage (%)[a]	14	14	14
Capex for LPG removal, dew point adjustment, and compressor, and trucks (US$/mmscf/d)[b]	8,750,000	4,250,000	3,700,000
Capex for LPG removal, dew point adjustment, compressor, and trucks (US$)[c]	6,125,000	14,025,000	24,420,000
Opex (as % of capex)	13	13	13
Equipment availability (% of year)	94	94	94
CNG price (US$/mscf)	6.00	6.00	6.00
AG price paid to operator (US$/mscf)	0.25	0.25	0.25
LPG price (US$/mscf)	3.00	3.00	3.00

Source: World Bank, based on data provided by GGFR.
Note: AG = associated gas; capex = capital expenditures; CNG = compressed natural gas; LPG = liquefied petroleum gas; mmscf/d = million standard cubic feet per day; mscf = thousand standard cubic feet; opex = operating expenditures.
a. Seven percent to power LPG extraction, 3 percent to run dew point adjustment, and 4 percent for compression to 3,000 pounds per square inch (psi).
b. Cost estimates to compress gas from 40 to 3,000 psi are based on 2020 data provided by compressor manufacturer Ariel, with an additional allowance for installation and civil works. For CNG transportation, it is assumed that standard 40-foot containers (capacity 212,000 scf) would be used. Cost estimates for CNG containers are based on actual 2020 costs for CNG containers sourced in China for a project in Nigeria.
c. Based on gas treated and compressed per day in year 3 of the project, approximately equivalent to the average across the seven-year project period.

annum). The model assumes a maximum distance to market of 150 km. Depending on the prevailing CNG prices in the target market, longer distances may be affordable; however, because the individual truck size is essentially fixed, increased distances rapidly increase the size of the fleet, which increases manageability and road safety issues. A CNG producer interviewed in Nigeria, for instance, indicated 300 km as the maximum radius of its operations, although at a higher CNG price than that assumed in this model.

The difference between the CNG price and the associated gas price paid to the oil field operator, as well as the ability to monetize LPG, is critical to the financial success of this FMR solution. The model assumes that the FMR developer buys raw associated gas from the oil producer and takes full charge of the LPG extraction and gas pretreatment before it is compressed into CNG. With such limited effort and investment required by the producer, it is plausible to pay very low prices for associated gas (US$0.25/mscf is assumed in this model). The model base case conservatively assumes a CNG price of US$6.00/mscf, resulting in a margin of US$5.75/mscf above the associated gas price.[8] As discussed in the next subsection, at this margin IRRs and NPVs are attractive, at least in medium and large flares. The sensitivity analysis includes a scenario in which the FMR developer is also able to keep and sell LPG at US$3.00/mscf, which further increases returns. In practice, how the LPG and CNG pie is split between the oil field operator and the FMR developer will likely be a subject of negotiation.

Financial returns and sensitivities

Under base case assumptions, the CNG solution is financially attractive only for medium and large flare sites (figure 3.22). The pretax, unlevered IRR of such investments is 16 percent in medium flares and 23 percent in large flares (figure 3.23), and NPVs are positive. Both IRR and NPV are substantially negative in the small flare scenario, reflecting the poor economies of scale at that flare size.

FIGURE 3.22

Compressed natural gas: Base case net present value sensitivity

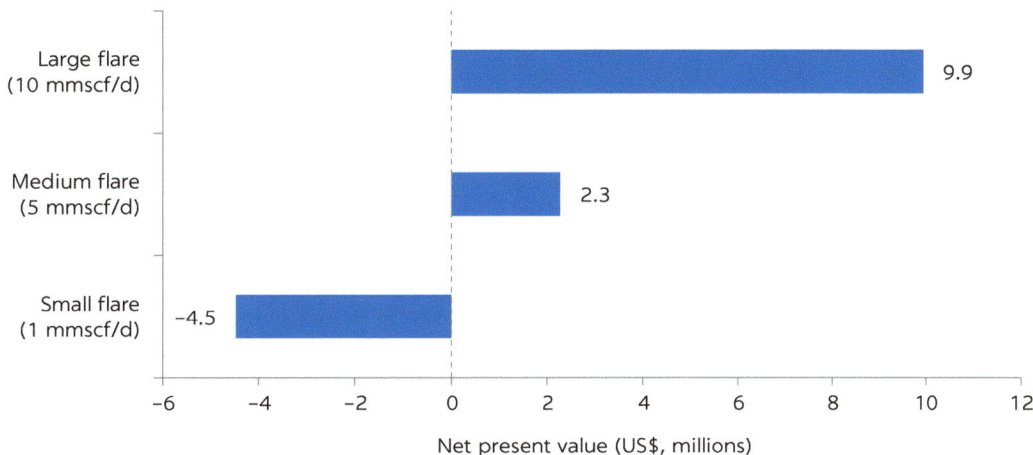

Source: World Bank.
Note: mmscf/d = million standard cubic feet per day.

FIGURE 3.23

Compressed natural gas: Base case internal rate of return sensitivity

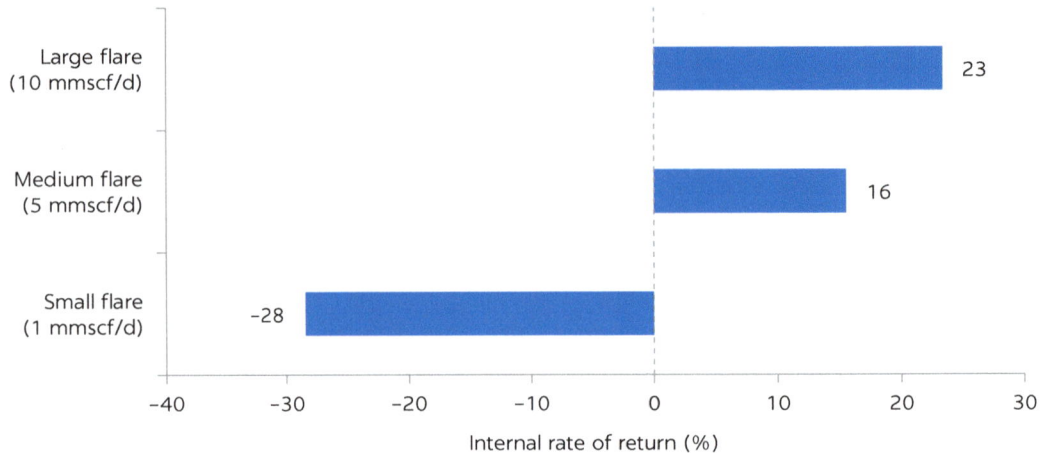

Source: World Bank.
Note: mmscf/d = million standard cubic feet per day.

The sensitivity analysis shows that, for CNG in small flares, IRR and NPV remain negative even when the model inputs are set at more favorable levels (figure 3.24). Focusing on medium and large flares, capex and CNG price are the variables that most affect returns. A 10 percent decrease in capex relative to the base case assumption adds 9–10 percentage points to the IRR, whereas a 10 percent increase causes IRRs to fall by 5–6 percentage points. A 10 percent increase in CNG price (to US$6.60/mscf) adds 8–10 percentage points to the IRR, whereas a 10 percent decrease (to US$5.40/mscf) cuts 6 percentage points from the base case IRR. Changes in project life and associated gas price cause more muted swings in IRR.

The sale of LPG by the FMR developer at US$3/mscf increases returns but without changing the picture substantially. Should the FMR developer obtain full ownership of the LPG and sell it at US$3.00/mscf, the IRR would increase by 3–4 percentage points in the medium and large flare scenarios.

SMALL-SCALE LIQUEFIED NATURAL GAS

Overview and assumptions

In this solution, modular liquefaction units are installed to convert associated gas into LNG, which is then transported via road trailers and sold (as LNG or gas after regasification; figure 3.25). The analysis assumes that (1) the heating value of the LNG to be delivered does not need to be controlled by removing the heavier components in the gas (if the energy content of the gas has to be controlled to meet the LNG buyer's requirements, then LPG extraction would be required prior to liquefaction); (2) the project is land-based and the LNG is transported by road trailer; (3) equipment required for use of the LNG by the LNG buyer (e.g. regasification, truck LNG fueling equipment) is for the buyer's cost; and (4) the FMR developer purchases raw associated gas and sells LNG (no tolling fee).

FIGURE 3.24

Compressed natural gas: Internal rate of return under different scenarios

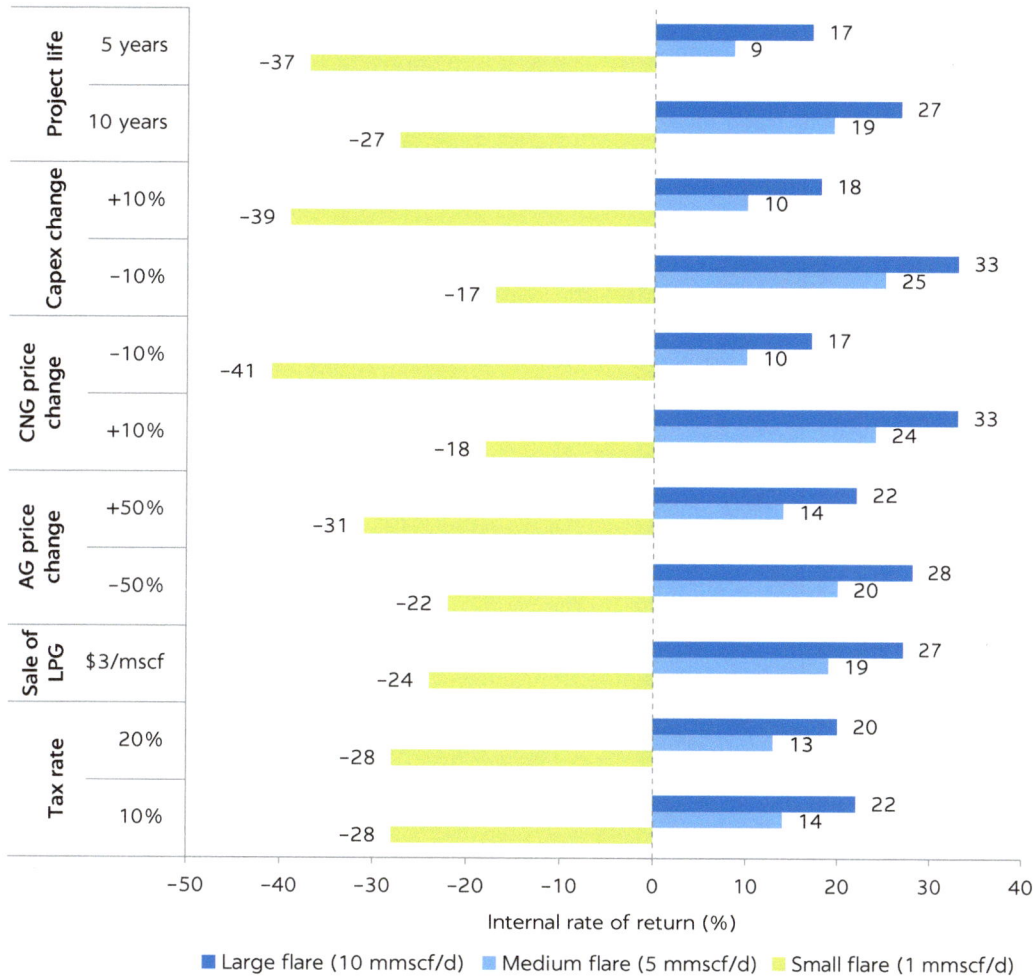

Source: World Bank.
Note: AG = associated gas; capex = capital expenditures; CNG = compressed natural gas; LPG = liquefied petroleum gas; mmscf/d = million standard cubic feet per day.

Capex in this solution benefits from significant economies of scale. The investment required for the pretreatment, liquefaction, and transportation of 1 mmscf/d of associated gas in the large flare scenario is less than 60 percent the investment on a US$ per mmscf basis required in the small flare scenario (US$5.8 million vs. US$10 million). As a result, although a large flare is 10 times as large as a small flare (10 mmscf/d vs. 1 mmscf/d), the required capex is just over 5 times as large.

Although not as large as in the CNG solution, opex is a substantial cost component in small-scale LNG as well. Because LNG has a higher energy density than CNG and therefore requires fewer trucks to transport the gas, the model assumes a maximum distance to market by trucks of 300 km. Annual opex is estimated at 8 percent of capex.

As for CNG, the margin between end-product price (LNG in this case) and associated gas is critical to financial returns. The model assumes that the FMR developer buys raw associated gas from the oil producer, paying a very low price of US$0.25/mscf. LNG is assumed to be sold at US$8.00/mscf, a higher price than for CNG reflecting the greater distance LNG can be economically

FIGURE 3.25
Schematics of small-scale liquefied natural gas solution

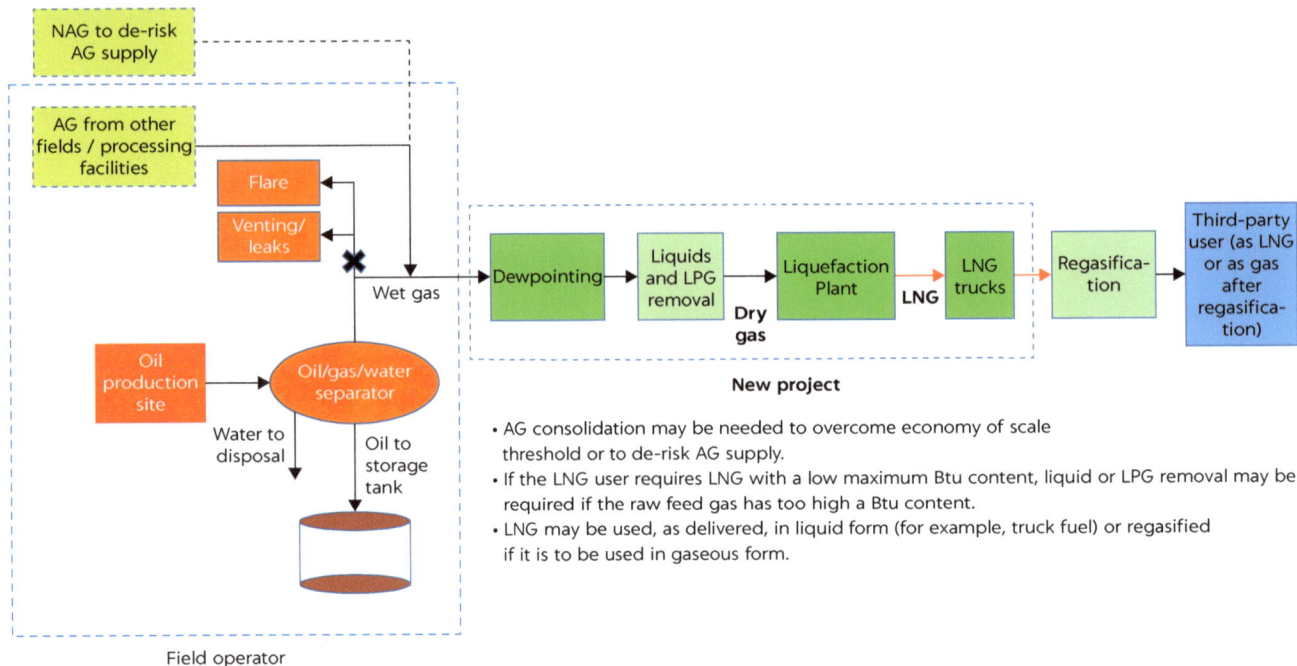

Source: World Bank.
Note: Nonassociated gas (NAG) is not essential in many projects. In some projects, however, NAG serves as a backup supply of gas, de-risking a project.
AG = associated gas; Btu = British thermal unit; LNG = liquefied natural gas; LPG = liquefied petroleum gas.

transported to find a market commanding this price. As discussed in the next subsection, at these LNG and associated gas prices, IRRs and NPVs are attractive, at least in medium and large flares. In practice, the oil field operator may want to extract a higher price for the associated gas if it senses that the FMR developer will make very high returns. Table 3.6 lists all assumptions specific to this solution.

Financial returns and sensitivities

Under base case assumptions, the small-scale LNG solution is financially attractive only for medium and large flare sites (figure 3.26). The pretax, unlevered IRR of such investments is 20 percent for medium flare sites and 24 percent for large flares, and NPVs are positive (figure 3.27). Both IRR and NPV are negative in the small flare scenario, reflecting the poor economies of scale at that flare size.

The sensitivity analysis shows that, for LNG in small flares, IRR and NPV remain negative even when the model's inputs are set at more favorable levels (figure 3.28). Focusing on medium and large flares, project life, capex and LNG price are the variables that most affect returns. Increasing the project life from 7 to 10 years adds 3–4 percentage points to the base case IRR; decreasing it to 5 years cuts 7 percentage points. A 10 percent decrease in capex relative to the base case assumption adds 4–5 percentage points to the IRR; a 10 percent increase

TABLE 3.6 **Assumptions regarding small-scale liquefied natural gas**

ASSUMPTIONS	SMALL FLARE (1 mmscf/d)	MEDIUM FLARE (5 mmscf/d)	LARGE FLARE (10 mmscf/d)
Fuel gas ratio (%)[a]	98	98	98
Parasitic gas usage (%)[b]	15	15	15
Capex for liquefaction and trucks (US$/mmscf/d)[c]	10,000,000	6,200,000	5,750,000
Capex for liquefaction and trucks (US$)[d]	8,000,000	22,320,000	41,400,000
Opex (as % of capex)[e]	8	8	8
Equipment availability (% of year)	98	98	98
LNG price (US$/mscf)	8.00	8.00	8.00
AG price paid to operator (US$/mscf)	0.25	0.25	0.25

Source: World Bank, based on data provided by GGFR.
Note: AG = associated gas; capex = capital expenditures; LNG = liquefied natural gas; mmscf/d = million standard cubic feet per day; mscf = thousand standard cubic feet; opex = operating expenditures.
a. Assuming no LPG extraction.
b. 3 percent to run dew point adjustment, 12 percent for liquefaction.
c. LNG Liquefaction capital costs are based on 2020 data from manufacturers Western Shell Cryogenic Equipment Co. and Siemens for modular units ranging in size from 0.4 mmscf/d to 2.5 mmscf/d feed. An additional allowance for installation, utilities, and civil work costs has been included. For LNG transportation, it is assumed that standard 55 m³ trailers (equivalent to ~1.2 mmscf gas) would be used. The cost of a trailer is based on a 2020 quote provided by a Chinese manufacturer for application in Nigeria.
d. Based on gas available in year 3 of the project, approximately equivalent to the average across the seven-year project period.
e. For preprocessing, liquefaction, and delivery of the LNG by trailer to a maximum distance of 300 km.

FIGURE 3.26

Small-scale liquefied natural gas: Base case net present value sensitivity

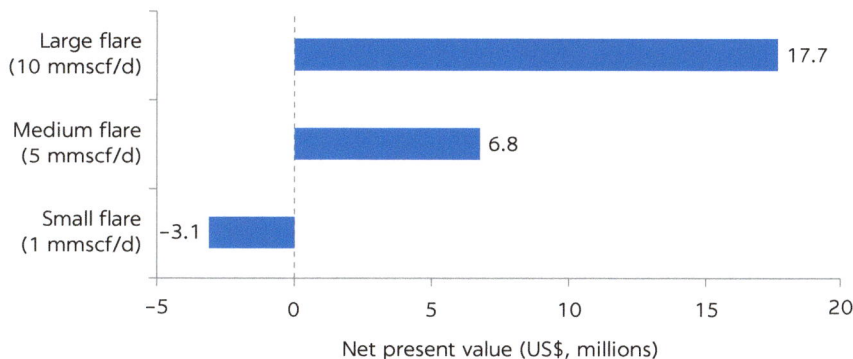

Source: World Bank.
Note: mmscf/d = million standard cubic feet per day.

FIGURE 3.27

Small-scale liquefied natural gas: Base case internal rate of return sensitivity

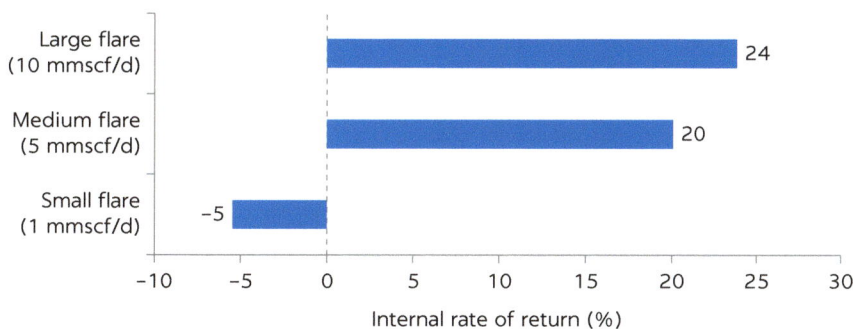

Source: World Bank.
Note: mmscf/d = million standard cubic feet per day.

FIGURE 3.28

Small-scale liquefied natural gas: Internal rate of return under different scenarios

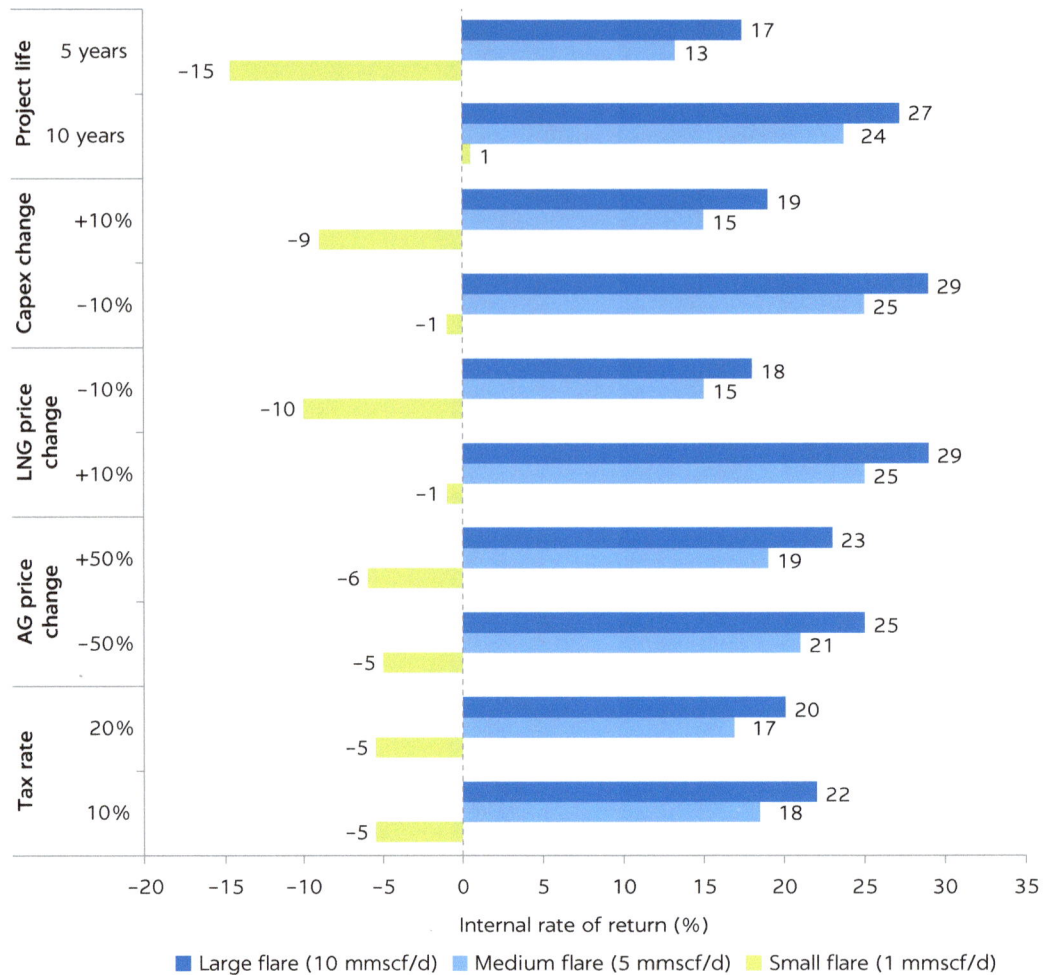

Large flare (10 mmscf/d) Medium flare (5 mmscf/d) Small flare (1 mmscf/d)

Source: World Bank.
Note: AG = associated gas; capex = capital expenditures; LNG = liquefied natural gas; mmscf/d = million standard cubic feet per day.

causes IRRs to fall by 5 percentage points. A 10 percent change in LNG price has a very similar impact on IRRs as the 10 percent change in capex. Changes in associated gas price, conversely, cause more muted swings in IRR.

NOTES

1. A single-digit IRR is below the minimum return threshold of 10 percent indicated by many industry participants.
2. It is also assumed that the developer has no other taxable income that can be consolidated with project income or losses for tax purposes.
3. Flares greater than 1 mmscf/d and less than 10 mmscf/d were globally 2,358, representing 48 percent of the total flared gas volume and 32 percent of the total number of flare sites.
4. Includes gas compression, power generation, and back-end connection.
5. Wholesale prices represent the floor, and FMR projects may be able to secure higher prices, for instance by targeting specific off-takers.

6. Typically, the credit rating of an oil company is significantly better than that of a developing country distributor. For that reason, it makes sense to sell at 0.08 US$/kWh to a distributor (with exchange rate risk) versus selling at 0.07 US$/kWh to an investment grade (US$ denominated) operator.

7. These figures are presented with no reference to countries, oil producers, and FMR project developers to comply with confidentiality agreements.

8. CNG may often be used for power generation, displacing diesel as a fuel. Because the international price of diesel has exceeded US$10/mmBtu for most of the past 13 years, CNG even at US$8/mmBtu would offer an attractive cost saving on fuel. For instance, a CNG developer in Nigeria indicated that CNG prices in that country are in the range of US$8.00/mmBtu to US$9.00/mmBtu as of April 2021.

4 Case Studies

This chapter presents six case studies of flaring and methane reduction (FMR) solutions. Five case studies cover actual projects, and one (Nigeria) describes a novel regulatory approach to FMR. The case study selection attempts to reflect the technical and geographic diversity of FMR projects. From a solutions standpoint, three case studies discuss gas-to-power projects (Aggreko, Hoerbiger, Mechero), one discusses a liquefied natural gas (LNG) project (Galileo), and one an innovative digital flare mitigation approach (Crusoe Energy). From a geographic standpoint, the case studies discuss projects and solutions applied in Latin America, the Middle East, Nigeria, North America, and the Russian Federation. The case studies were developed on the basis of interviews with company management, external experts, documents shared by the case study subjects, and publicly available information. Each case study ends with a few takeaways that complement the findings of the financial analysis in chapter 3.

SUMMARY TAKEAWAYS

The case studies offer a concrete perspective on the following key success factors affecting FMR projects:

- *The FMR developer's ability to provide turnkey solutions and execute swiftly.* FMR projects are seen by oil companies as unworthy diversions of capital and engineering resources. Such concern is especially the case at small flare sites, for which even the avoidance of flare fines may not be a sufficient incentive for the operators to directly engage in FMR. In these cases, FMR projects are more likely to be executed when FMR developers can take care of the whole problem on behalf of oil companies—including design, procurement of equipment (or use of developer-owned equipment), installation, financing, and operation and maintenance. In response, Aggreko, Galileo, and Hoerbiger have all transitioned from traditional equipment rental (Aggreko) or supply (Galileo and Hoerbiger) to the provision of turnkey project management solutions under long-term contractual arrangements.

- *Modularity and movability of equipment.* Similar to the future production profile of an oil field, the exact magnitude, longevity, decline rate, and chemical composition of an associated gas flare are never known with certainty up front. Successful FMR developers address this risk by deploying modular equipment that can be easily down- or upsized during the course of a project and shifted among different flares within a reasonable geographic range. Tackling a portfolio of flares rather than an individual one is an inherent hedging strategy. Aggreko, Crusoe, Galileo, and Hoerbiger all stressed the importance of modular equipment and the ability to redeploy quickly from site to site.

- *The FMR developer's ability to finance a project solely or primarily through equity.* An FMR developer's capacity to equity-finance—wholly or in large part—the construction phase of the project provides certainty of execution to the oil company that commissions a project. Conversely, projects whose financing is reliant on securing bank loans are potentially exposed to delays or, ultimately, the risk of not reaching financial closure. This risk is material in FMR projects, whose features are in most cases not suitable for project finance, particularly because of (1) the uncertainty of associated gas supply and the resulting unwillingness of oil producers to enter into deliver-or-pay agreements;[1] (2) end-product (power or gas) price risk, especially when the remuneration of the FMR developer cannot be structured as a tolling fee; (3) the risk that the off-taker (for example, power distribution company) may not honor its contractual obligations; (4) project execution risks; and (5) macroeconomic risks such as currency fluctuations.[2] Aggreko, Galileo, and Hoerbiger rely primarily on equity to fund their FMR projects. Mechero managed to obtain loans for the FMR project analyzed in the case study, but the leverage ratio is lower than in a typical project finance structure. Crusoe has obtained secured loans against some of the equipment it uses.

- *Strong project management and execution capabilities.* Although present in all infrastructure-type projects, this risk is heightened in the FMR arena by the variety of stakeholders involved (oil company, off-taker, regulator, and local communities, among others) and the geographic dispersion of flares within a site. All FMR developers interviewed attributed their success in part to speed and quality of execution.

AGGREKO

Background

Aggreko is a supplier of power generation and temperature control equipment and is headquartered in Glasgow, Scotland, United Kingdom. The company is listed on the London Stock Exchange and has global operations across several sectors, including oil and gas.

Aggreko provides full turnkey solutions to operators of flaring fields, including the installation of Aggreko-owned modular gas-fired generators, equipment for the pretreatment of gas if needed (such as hydrogen sulfide [H_2S] treatment and natural gas liquids [NGLs]—stripping), emissions controls, and possibly participation in the infrastructure investment required for the full implementation of the FMR project.

Aggreko bears the capital and operating costs of the FMR projects, in exchange for the revenues realized from the sale of electricity over a contractual period. In most cases, electricity is sold back to the field operator for drilling, production, and other on-site operations, with prices that generally provide a considerable saving relative to diesel generation or that are competitive with grid prices (if the operator has access to it). In a few cases, instead of or in addition to on-field use, Aggreko sells power to existing grids. Figure 4.1 show Aggreko's installed capacity by power use, and figure 4.2 shows its geographical distribution.

Going forward, Aggreko is exploring the potential to sell electricity to business and industrial users, including power-hungry businesses such as data centers and cryptocurrency miners. As of this writing, Aggreko has signed its first contract with Bitcoin miners in the United States. Operators that outsource power production for own use to Aggreko do not need to engage in what they consider a noncore activity and avoid the up-front cost of installing gas turbines (plus any associated financing costs).

The following gas-to-power solutions refer to projects carried out by Aggreko in Russia, where existing regulation requires oil companies to ensure a minimum 95 percent use of associated gas. Lack of compliance entails the payment of substantive fines.

The Suzunskoye Field

The Suzunskoye Field is an oil field in Western Siberia (Russia), currently operated by RN-Vankor LLC, a subsidiary of PJSC NK Rosneft. The field's estimated flare is more than 60 million cubic meters (m^3) of flared associated gas per year.

The field operator had dual objectives: to eliminate gas flaring and to build a reliable source of power generation to support the operation of its 5.2 million

Aggreko's installed capacity by power use
Percent of total, number of projects

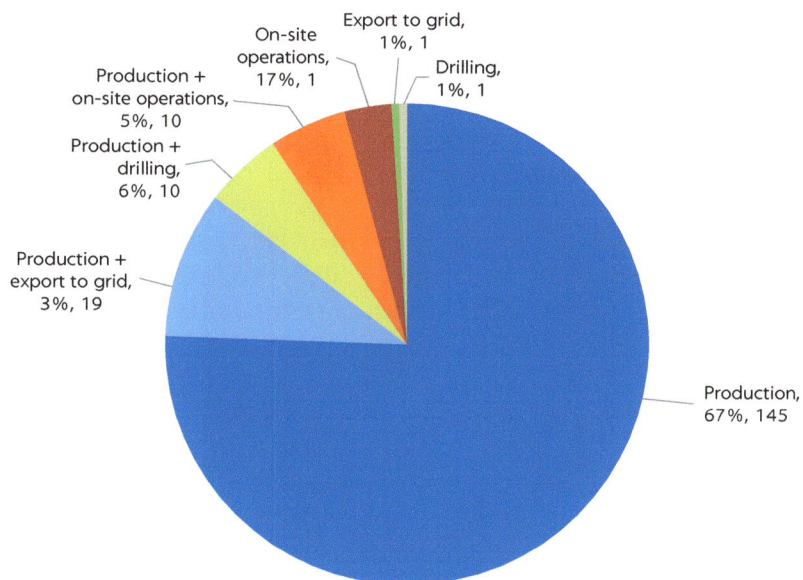

Source: World Bank, based on Aggreko's commercial literature.

FIGURE 4.2

Aggreko's installed capacity, by geography

Percent of total, number of projects

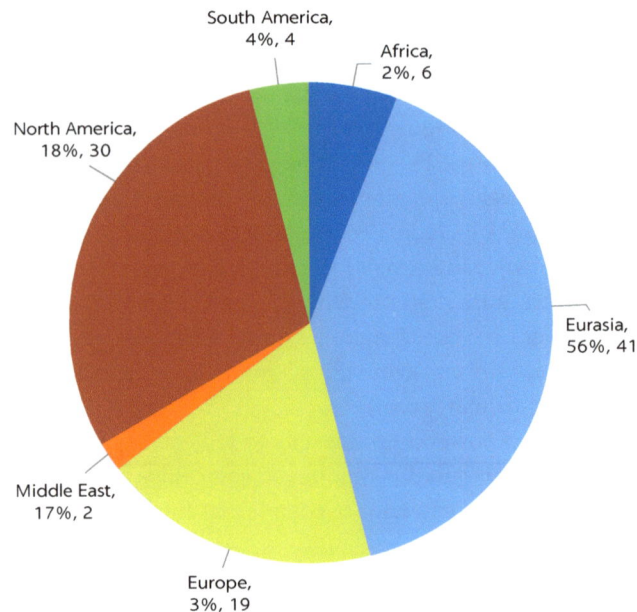

Source: World Bank, based on Aggreko's commercial literature.

tons per day oil treatment plant. To address these objectives, Aggreko proposed a long-term contract with 58 megavolt-amperes of total power capacity, which assumed supplying all necessary spare parts and materials, as well as performance of scheduled technical works. Equipment includes eight transformers with a unit capacity of 6,300 amperes, two fuel tanks of 20,000 liters each, one load bank of 1,266 kilovolt-amperes, and 13 kilometers (km) of cable to ensure a stable power supply during the drilling and operation of oil wells. The project's capital expenditures (capex) and operating expenditures (opex) are assumed by Aggreko.

The project is an extension of a smaller project that began in 2016 and grew in scope over time. Aggreko is responsible for ensuring commissioning and round-the-clock maintenance of the generator sets and utility networks. The company delivered equipment on-site and installed and commissioned it within 30 days from the date of signing the contract, which enabled the field operator to obtain power for progressive production of raw materials, start work promptly, and operate the field one year earlier than initially planned. The remote location of the field and freezing temperatures (which required upgrades to normal winterization equipment) were part of the technical challenges.

The project was initially designed to work on natural gas supplied at high pressure, so Aggreko built a pressure reduction unit. The plant was later modified to allow working directly on associated petroleum gas supplied at normal pressure, thus providing fuel flexibility to the operator.

The project is estimated to achieve more than 120,000 tons of carbon dioxide (CO_2) emissions reduction per year and is securing supply of power to the operator.

The Yuzhno-Priobskoye Field

The Yuzhno-Priobskoye Field in Russia is an oil field in Western Siberia and is currently operated by Gazpromneft-Khantos LLC, a subsidiary of Gazprom Neft. The field's estimated flare is more than 30 million m³ of associated gas per year.

The operator had a dual objective: to reduce flaring and to reduce the cost of power generation needed for upstream facilities, which usually account for up to 40 percent of production costs. To address these objectives, Aggreko supplied 24 megawatts (MW) with a significant guaranteed discount to grid price level as the turnkey project, without capex investment for the operator. The development and commissioning works were carried out by Aggreko, which will also provide maintenance services for the power plant throughout the seven years of the contract term.

The project was initially designed to work on stripped gas but, after an additional request from the customer, was later upgraded to allow consumption of associated petroleum gas, thus providing fuel flexibility to the operator.

By selling part of the associated gas to Aggreko, Gazpromneft-Khantos addresses the need to eliminate flaring at the Yuzhno-Priobskoye Field. The power plant operates in parallel with the grid, thus ensuring reliability of power supply.

This project is estimated to achieve more than 60,000 tons of CO_2 emission reduction per year and is securing the supply of power with competitive prices.

The Sredne-Nazymskoe Field

The Sredne-Nazymskoe Field is an oil field located in Western Siberia, Russia. The field is currently operated by RITEK LLC, a subsidiary of PJSC Lukoil. The field's estimated flare is about 3 million m³ of associated natural gas per year. The field's operator had a dual objective: to eliminate flaring and to reduce the cost of power needed for its operations by replacing diesel with the less-polluting natural gas.

To address these objectives, Aggreko built a power plant with 8.7 megavolt-amperes of gas power capacity, two transformer substations, and one switchgear. An additional feature of the solution is a super capacitor, which allows for stabilization of the frequency and voltage in the network in case of sudden changes and fluctuations in electrical power consumption required for drilling operations. The operator's up-front cash outlay was limited to mobilization cost, and Aggreko ensured all capex and opex associated with the project.

By eliminating flaring, the project is estimated to achieve more than 5,000 tons of CO_2 emission reduction per year and is securing the supply of power to the operator. Additional emissions reductions due to fuel switching were not quantified. Power cost savings were estimated at 50 percent on account of fuel switching.

Middle East Project

Aggreko is engaged in an FMR project in the Middle East that is operated by an international oil major. The field is estimated to flare about 1.1 million m³ in years 1 and 2, possibly reducing thereafter. The gas requires pretreatment and NGL stripping. Because of emissions limits and potential fines for noncompliance, the field's operator was unable to expand production over 30,000 barrels per day because such expansion would have resulted in additional flaring. Furthermore, the operator wanted to maximize the use of available gas for power by the local government. The operator thus had an overarching objective to increase production and deliver power at competitive rates to the local grid.

Aggreko developed and financed the project from concept to mobilization, commissioning, and operation to deliver a turnkey solution for the operator. The fleet and infrastructure employed are worth over US$50 million. The project required a new 6 km pipeline from the flare stack to the Aggreko site, along with full gas treatment and an NGL production plant, four 50 megavolt-ampere transformers, switchgear, 7 km of new 33 kilovolt (kV) overhead transmission lines, and an upgrade to an existing 33km of 132 kV transmission lines—all within Aggreko's scope.

Aggreko installed 150 MW of power generation capacity based on its 1 MW mobile and modular gensets, which were used because of their ease of mobilizing and demobilizing, and low derating factor at high temperatures. At the time of writing, 100 MW is already operating. The gas pipe, treatment infrastructure, and power connection will be transferred to the customer at the end of the project; but all power assets remain Aggreko's property and will be demobilized at the project's conclusion or in stages as gas volumes deplete.

Given regional instability, specific regulatory challenges in the country, and the difficulty of enforcing contracts, it was important for Aggreko to engage in tri-party discussions with the operator and utility operator. The project helped the customer offset the need for additional capacity by saving 840 tons of CO_2 per day. Socioeconomic cobenefits also resulted from the project, such as delivering 150 MW of power onto the local grid, stabilizing supply, reducing cost, and reducing the local dependency on expensive high-emissions liquid fuel plants.

Takeaways

Oil companies value the ability of FMR developers to provide turnkey solutions in a noncore activity such as flare reduction. Aggreko owns the power generation equipment it installs and puts its balance sheet to work to build necessary infrastructure (gas treatment, gas connection, overhead power lines)— eliminating the need for capital investment on the part of the oil company. To support its business model, Aggreko has set up a specialist gas-to-power division to lead design, engineering, operations and maintenance, and, where appropriate, the sale of power to third parties. The operator obtains power at a discount relative to diesel generation or, depending on the location, expensive power from the grid (as is the case in Western Siberia).

Speed of deployment and the flexibility of the equipment installed can be critical in persuading oil companies to outsource FMR to independent developers that can mobilize their fleet of generation sets quickly, and also to remote locations. The use of modular equipment allows FMR projects to expand or shrink, seconding the production dynamics of the field and associated gas as well as the level of power demand from the operator or third-party off-takers where applicable. This aspect is important because volume and specification of gas are rarely known at the project design phase, and they can change over the life of the field.

HOERBIGER

Background

Hoerbiger is an Austrian original component manufacturer (OCM) for gas compressor sealing elements and related equipment that expanded into turnkey FMR project development. The company had revenues of €1.1 billion in 2019,[3] approximately 50 percent of which were derived from the strategic compression technology business unit in the oil and gas industry and the remainder from automotive components.[4]

Hoerbiger started developing FMR gas-to-power projects in Ecuador and is now looking to expand to other Latin American countries. Hoerbiger's FMR portfolio in Ecuador includes 11 projects with investment sizes in the US$2 million to US$6 million range. This case study focuses on Hoerbiger's FMR project in the Sacha Central oil field in Ecuador. Ecuador is a fertile environment for independent FMR project developers for several reasons: (1) the national oil company EP Petroecuador is unable to fund FMR projects because of budgetary constraints at current oil prices, and the preference to direct investment to core oil production operations; (2) the Ecuadorian oil and gas industry relies primarily on expensive diesel generation for its operational needs, with diesel being imported because of limited oil refining capacity in the country; (3) the Ecuadorian regulatory environment is relatively easy to navigate compared to other Latin American countries; and (4) the Ecuadorian government is committed to Zero Routine Flaring by 2030.[5]

Hoerbiger's Sacha Central Project

The Sacha Central FMR project is a gas-to-power (on-site use) project. It uses 1 million standard cubic feet per day (mmscf/d) of associated gas from a 1.5–2.0 mmscf/d flare in a field operated by Petroecuador. The project converts associated gas into electricity that is delivered to Petroecuador's captive grid, powering the company's operations at an agreed-on tolling fee. The project started operations in April 2021, after a six-month design and construction process. Hoerbiger's established relationship with Petroecuador contributed to a swift development schedule.

The project was developed jointly by Hoerbiger and Arcolands, the official distributor of Ineos's Waukesha and Jenbacher equipment (Waukesha gas engines and generators) in Ecuador. Both companies have shifted their business model to the provision of turnkey solutions. Instead of selling (or reselling, in the case of Arcolands) equipment for an up-front payment, they fund the initial investment; install, operate, and maintain the equipment; and recover their costs plus a return through a multiyear tolling agreement.[6] No special purpose vehicle was set up in this project. Arcolands was engaged as contractor and Hoerbiger as subcontractor. Hoerbiger was in charge of engineering, procurement, and construction (EPC) of the balance of plant, which encompasses all infrastructure (for example, foundations, piping and pipelines, electrical installation) needed to accommodate all necessary equipment like compressors, gas engines/generators, and electrical interchanges. Arcolands provided the power generation equipment. After completion of the project, ongoing maintenance contracts (Arcolands for power generators, Hoerbiger for compression equipment) secured the availability of the installed equipment. Figure 4.3 is a schematic representation of the Sacha Central flare-to-power business model.

The project required an investment of US$5 million for power generators, gas compressors, electrical interchange, and the required infrastructure. Four MW of power capacity was installed. The investment was 100 percent equity-financed by Hoerbiger and Arcolands (US$2.2 million and US$ 2.8 million, respectively).

The project's revenue model is a tolling agreement with monthly capacity and, to a lesser degree, energy-related payments. The capacity payments are due regardless of the availability of associated gas and also in force majeure situations (such as employee strikes). Payments are not due only in the scenario in which Hoerbiger is unable to provide the service owing to failure of its equipment. The tolling agreement has a duration of four years. Thereafter, Hoerbiger receives only operations and maintenance fees but no capacity payments. Petroecuador's commitment under the tolling agreement is not covered by any guarantees—Hoerbiger assumed the full risk.

The implied cost of electricity during the four-year contract is competitive compared to diesel generation and still sufficiently high to generate an estimated 10 percent return on equity for Hoerbiger. The tolling agreement will result in US$10 million in payments to Hoerbiger over four years, equivalent to US$0.10 per kilowatt-hour (kWh) of power supplied to Petroecuador. Because capacity payments are not applicable after year 4, over a longer period the average electricity price drops further. This price compares to a cost of diesel generation for Petroecuador currently ranging between US$0.20/kWh and US$0.26/kWh (assuming a price of diesel in the range of US$2.0 to US$2.5/gallon).[7]

In addition to reducing emissions from flaring and displacing diesel generation, the project has important socioeconomic cobenefits. Hoerbiger estimates

FIGURE 4.3

Sacha Central flare-to-power business model

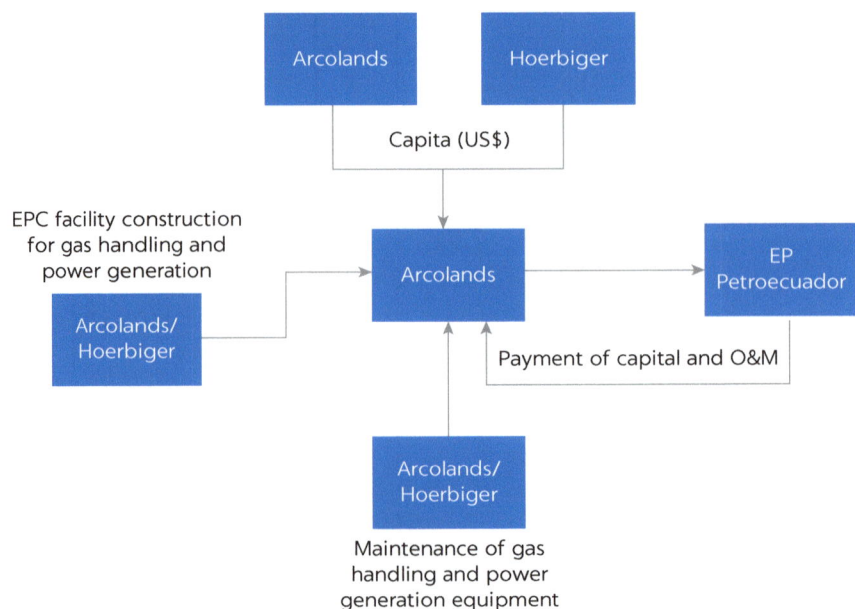

Source: World Bank, based on interviews with Hoerbiger.
Note: EPC = engineering, procurement, and construction; O&M = operations and maintenance.

the direct emissions savings from flaring at 70,000 tons of CO_2 equivalent (tCO_2e) per year over the four-year contract; further savings from diesel replacement were not quantified. From a social perspective, construction and operation generated employment and service supply opportunities for the local indigenous population. The project also contributed to addressing concerns raised by the local community, who attributed an increased incidence of health ailments to their proximity to the oil field (Colectivo 2020; *El Comercio* 2021).

Takeaways

Turnkey FMR project development can represent a new revenue opportunity for OCMs facing slowing demand for equipment in a mature oil and gas sector. In addition to their technical expertise, OCMs can put their balance sheet to work and relieve oil field operators of the up-front investment cost of an FMR project. The OCMs' familiarity with their client's operations can also contribute to speedy project development.

Hoerbiger attributes the successful implementation of Sacha Central and similar FMR projects to several factors: (1) strong project management and execution capabilities (construction in four months and within budget); (2) the small size of the projects, which inherently reduces execution risk; (3) Hoerbiger's and Arcolands' existing technical know-how as OCM and equipment reseller, respectively; (4) the small project size that did not warrant political interest groups to get involved; and (5) the clear climate change mitigation benefits of its solutions.

Building a portfolio of small FMR projects reduces execution and financial risks for developers. Small projects can be easier to implement, especially if the involvement of stakeholders is limited to the developer, oil company, and off-taker to avoid political risk. A portfolio approach also acts as a hedge against flare-specific risks, such as the unpredictability of future associated gas volumes.

Although the cash flow profile of some FMR projects could be suitable for a leveraged capital structure, often 100 percent equity funding is the only practical solution in the project development phase. Minimizing project development time and hassle is often critical to convincing oil field operators to consider FMR projects, because the latter are seen as noncore and bring limited financial benefits compared to oil production. Implementation and enforcement of greenhouse gas regulations increase interest and awareness in FMR projects as well, which is not always the case across countries. Negotiating loan packages can be time-consuming given the many uncertainties of FMR projects (for example, gas supply and off-taker risks), and the prospect of receiving a loan in the construction phase can be far from certain.

In the operational phase, when development risk is no longer present, FMR projects may be easier to leverage—especially if de-risking tools are available—freeing up equity that developers could dedicate to new FMR projects. Not all risks disappear in the operational phase; for instance, FMR projects remain exposed to off-taker risk. These risks will likely be reflected in relatively low leverage and relatively high interest payments. De-risking tools such as partial credit guarantees or insurance for breach of contract could enhance the bankability of FMR projects, allowing developers to take money off the table and subsequently spread their equity capital across a larger number of FMR projects.

MECHERO ENERGY

Background

The Floreña Field in Colombia is an oil field that was originally discovered in the early 2000s and is currently operated by Ecopetrol, an integrated oil and gas company headquartered in Bogotá. Colombia has restrictions and fines for gas flaring, which puts the onus on exploration and production companies to reduce flaring. Within this context, companies have tried to address the issue with gas-to-power projects and reinjection of the associated gas.

Operators at the Floreña Field in Colombia did not have a straightforward way to sell or dispose of the associated gas in their operations. Without off-take for the associated gas, it was difficult to maintain oil production from the field without incurring fines. There was investment in high-pressure reinjection compressors, but they limited output. Furthermore, flaring a quantity of associated gas that exceeded what could be reinjected would result in penalties that damage the economics of the field. This consideration was the driving force for a gas flaring reduction project.

Mechero Energy is a holding company based in the United States that develops energy-conversion solutions and aims to reduce pollution and increase efficiency around natural resource use. Mechero focuses not only on flaring but also on developing solutions within LNG, natural gas treatment plants, and power generation. Mechero develops and sets up special purpose vehicles for projects, including equity participation, and has commissioned three FMR projects in Colombia and Ecuador.

Mechero Energy is undertaking the Termo Mechero Morro (TMM) project, an FMR initiative to address the flaring issue at the Floreña Field. The TMM project started development in 2014 and has been operating since 2018. In total, operating the field would result in a flare size, net of reinjection, of about 50 mmscf/d. Other off-takers, a gas-to-power project led by a competitor of Mechero and a pipeline to a local town, use about 40 mmscf/d; and Mechero handles the remaining 10 mmscf/d to 12 mmscf/d. Any leftover associated gas is reinjected.

Mechero's gas-to-power solution

With Mechero as the sole developer of the project, the technical solution consists of three power plants that are fueled by the associated gas of the field and then send power to the regional grid. TMM uses three 19 MW power plants that are fueled by the associated gas of the field and deliver energy to the regional transmission system at 115 kV. Each of the plants consists of two Wartsila 20V34SG motor generators, whose nominal capacity is 9.3 MWe (megawatt equivalent) and a GE Jenbacher JGC motor generator with a capacity of 1.0 MWe (photo 4.1). The project includes the EPC of a transmission line (18 km in high voltage, 13.8 kV/115 kV), a 6-inch carbon steel gas pipeline of 2 km, and a system of gas conditioning by mechanical refrigeration to recover condensates (NGLs) and maximize the power of the plant. Building the transmission line entailed negotiating rights of way with 32 landowners. Mechero also built and donated a small substation to a nearby community.

TMM required an investment of about US$72 million, funded with equity and debt investment from domestic and international financial institutions.

Termo Mechero Morro

Source: © Mechero Energy. Used with the permission of Mechero Energy. Further permission required for reuse.

The funding was 45 percent equity and 55 percent debt. Equity was provided by Mechero and Ashmore Group, a fund manager based in the United Kingdom. Bancolombia and Davivienda, two Colombian financial institutions, provided project finance debt to complete the capex funding. Capex was about US$1.2 million per MW, which is a good price when compared to similar projects. The total investment of US$72 million includes power generators, front-end capex (pipeline from flare site to the power generation site), and back-end capex (16.5 km of transmission line and substations).

The project faced hurdles in securing capex financing in both equity and debt dimensions, as well as structuring the gas supply agreement. A private equity investor does not typically assume development risk, but Ashmore did in this case even though the project was not EPC-ready at the time of investment. It is otherwise a high bar to achieve to secure private equity financing for projects like TMM. On the debt side, the banks are lending only until 2024, because the gas supply agreement expires in 2025. Mechero had to secure a deliver-or-pay clause in the gas supply agreement to secure the banks' loans. Because the power market in Colombia is open and competitive, it was also important to the banks that, in order to ensure reliability, the off-taker was a large state-owned power generation company. Finally, in the gas supply agreement, Mechero had to accept a high associated gas price[8] because of the country's gas supply dynamics: Colombia is short of gas and is in a position that it must import LNG (until further discoveries relieve the supply situation).

The project achieved a good power purchase agreement price, but currency fluctuations and project delays are headwinds to achieving internal rate of return and net present value expectations on time. However, the gas supply agreement will need to be extended to achieve internal rate of return and net present value expectations, which are not achievable at the current rate because of the depreciation of the Colombian peso and project-related delays. To achieve that target, including a small residual value for the equipment at the end of the project,

Mechero will need to run the project until 2030—which adds an additional five years beyond the original end date. Further headwinds from the input side include the high associated gas price under the gas supply agreement and unhedged exposure to currency risk.

Takeaways

This case study highlights the challenges in achieving attractive internal rates of return and net present value in some FMR projects. Specifically, while negotiating good power prices in the power purchase agreement (in the context of the highly competitive Colombian electricity market), Mechero had to deal with high associated gas prices (the result of tight Colombian gas supply), unhedged local currency exposure, and project delays. The project is also exposed to the risk of renewal of the gas supply agreement in 2025—both well within the targeted 2030 project completion date. This risk explains, in part, why debt was a relatively low 55 percent of the capital structure.

The foreign exchange risk is challenging in Colombia and hedging long-term is not viable. There is a currency mismatch because all electricity in Colombia is sold in pesos and all gas (included associated gas) is sold in US dollars. The currency risk is very large, and long-term hedges are not available in Colombia because of the high volatility of the currency. The project's debt is in pesos, so creditors therefore take the risk that a sudden depreciation of the peso negatively affects cash flows as the peso-equivalent cost of associated gas rises significantly. For example, the US dollar to Colombian peso rate was 2,902 at the start of 2018 (when the TMM project began) compared to 3,742 in May 2021.

The existing mismatch between developer and private equity investor funding thresholds in FMR projects can deprive otherwise bankable projects of the necessary financing to proceed. Private equity investors prefer that projects be EPC-ready and all contracts be in place before investing. However, developers may not have enough capital to achieve EPC-ready status, and thus projects do not move forward. In the case of the TMM project, Ashmore assumed development risk, which is not typical of private equity investors in FMR projects in Colombia.

GALILEO

Background

Galileo Technologies is a manufacturer of modular technologies for the production and transportation of liquefied and compressed natural gas. The company was founded in 1987 and exports 90 percent of its products to clients in 70 countries.[9] Galileo is partly owned by Blue Water Energy, a midmarket private equity firm based in London and specialized in the energy sector.

One of Galileo's core proprietary technologies is the Cryobox, a portable, plug-and-play LNG production station capable of producing approximately 15 tons of LNG per day. The size of a 40-foot container, a portable Cryobox can be delivered in a single trail and easily relocated. Galileo also manufactures a mobile Cryobox-Trailer that can be used in virtual pipeline applications. Installation requires only concrete pads for the Cryobox erection and connection to gas sources, electricity, compressed air, and internet. Galileo also manufactures gas conditioning units to remove liquids and impurities from gas before injection into the Cryobox (one unit can upgrade natural gas for up to four Cryobox units). According to the company, Cryobox and conditioning units can be installed in as little as six months, compared to an average of two years for mini-LNG plants.

LNG produced by Galileo's Cryoboxes is transported on ISO (International Organization for Standardization) containers and used, after regasification, for a variety of applications, including power production, heating of industrial facilities, and other industrial processes. On average, LNG can be stored in standard ISO containers for up to three months,[10] allowing it to be transported potentially very long distances (hundreds or thousands of kilometers). Galileo also manufactures regasification units and has started including in its turnkey package 3.5–4.0 MW gas turbines supplied by the leading manufacturers in the market.

Galileo's FMR solution

Because of its modularity and portability, Galileo's technology is particularly suited to use in FMR projects. Cryoboxes are currently in operation at flare sites in Argentina, Australia, Colombia, and the United States. Galileo will soon deploy another 34 Cryoboxes in Brazil. The minimum flare size required to install a Cryobox is 1 mmscf/d, of which 0.8 mmscf/d on average is converted into LNG and the remainder is used to power the Cryobox. Multiple Cryoboxes can be installed to use associated gas from large flares or from clusters of small flares. The equipment can be easily redeployed to different locations, seconding flare patterns at a given field. A Cryobox has a very long useful life (decades), which allows Galileo to amortize equipment across multiple projects over time. Galileo estimates that one Cryobox reduces emissions from flaring by 13,000 tCO_2e per year, converting 5,000 tons of associated gas into LNG and also eliminating methane emissions from incomplete flaring at the same time.

In FMR projects, LNG produced with Galileo's equipment is usually delivered to the oilfield operator as anchor customer; the remainder is sold in the market, for instance to industrial users, gas station operators, and independent power producers. The operator uses LNG (after regasification) for captive power production. Once regasified, LNG from one Cryobox can support 3 MW per day of power generation capacity. This model is particularly applicable in oil basins where power production is centralized in a given location and electricity is transmitted via power lines to individual wells. In Argentina, for instance, the average distance between wells and the power generation unit is 700 km. In Galileo's Brazil project, the well is 1,300 km away from the power plant. Captive power production from LNG is currently competitive with diesel generation—the default power source for oil operators—in the oil basins targeted by Galileo. For example, Galileo's ZPTS upgrading/treatment module and Cryobox liquefaction project in North Dakota comprise one Cryobox that captures 1.1 million standard cubic feet per day (mmscf/d) of previously flared gas from six wells, and converts it into 0.7 mmscf/d (14 tons per day) of LNG. Galileo's gas treatment solution is also installed, capturing 0.2 mmscf/d worth of natural gas liquids.

Galileo relies on three revenue models for its FMR projects, depending on client preferences:

1. *Straight OEM (original equipment manufacturer) model.* In this scenario, Galileo sells its equipment, including installation and operations and maintenance, to the FMR project developer.

2. *Service fee model.* In this scenario, Galileo installs its equipment and is remunerated with fees for the provision of turnkey services, from the treatment and liquefaction of associated gas to the transportation and delivery of LNG. The typical contract duration is 5–10 years. The fee structure consists of a fixed amount per month based on installed capacity plus a processing fee for the LNG produced.

3. *LNG sale model.* In this scenario, a special purpose vehicle is set up to run the entire process. The special purpose vehicle purchases associated gas from the oil producer and sells LNG or electricity—if Galileo also installs power generation units—to the oil producer and third parties. Depending on associated gas quality, some projects are also able to monetize liquids by selling them to third parties. Projects are equity-funded by Galileo and, case by case, by co-investors such as public or private utilities and other companies. This model, which is asset-light from the perspective of the oil operator and entails the provision of turnkey services by Galileo, is gaining traction among clients.

Takeaways

Galileo attributes the success of its FMR solution to several factors: the modularity and movability of its technology, the ability to provide turnkey services to oil operators, the ability to shoulder the up-front investment instead of the oil operator, and swift project execution.

In addition to reducing emissions from flaring, Galileo's solutions also eliminate methane emissions from incomplete flares and leave no environmental footprint. Galileo's equipment can be installed without permanent damage to the ground and, once removed, allows for the quick return of native flora and fauna.

THE NIGERIAN GAS FLARE COMMERCIALISATION PROGRAMME

Background

Despite successful efforts by the Nigerian government to curb flaring over the past 20 years, Nigeria was still responsible for 5 percent of gas flared globally in 2020. That year, the Nigerian oil industry flared 7 billion cubic meters of associated gas, a 70 percent reduction over the previous 20 years achieved primarily by tackling large flares through own consumption and large revenue-generating FMR projects. In 2019—before the COVID-19 (Coronavirus) crisis—according to Nigeria's Department of Petroleum Resources (DPR), 178 sites in the country were flaring close to 900 mmscf/d, with some three-quarters of it being routine flaring (figure 4.4).

Despite various legal and regulatory actions since the 1960s, Nigeria did not witness a structural reduction in flaring until the turn of the century, coinciding with the development of the LNG industry and other gas use schemes. The Petroleum Act of 1969 gave ownership of all associated gas to the federal government free of cost at the flare and without payment of royalty. The Petroleum (Drilling and Production) Regulation of 1969 required producers to include associated gas use plans in field development plans for approval. The Associated Gas Re-injection (Continued Flaring of Gas) Regulations of 1985 required producers to reinject associated gas when a use scheme was not feasible and introduced

FIGURE 4.4

Flared gas volume in Nigeria, 1992–2019

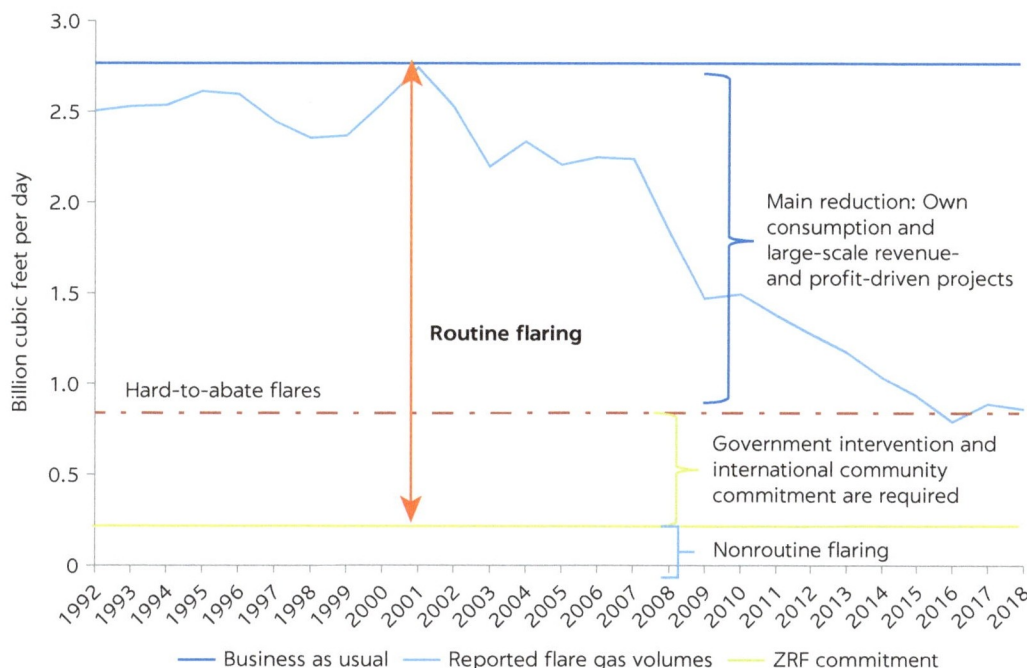

Source: World Bank, based on data provided by the Nigeria Department of Natural Resources.
Note: ZRF = Zero Routine Flaring.

flare payments. Despite these measures, gas flaring saw no structural reduction in the last century—a slowdown in flaring in the 1980s simply mirrored the slowdown in oil production at the time. The launch of Nigeria's first LNG plant in 1999 and the development of various gas use schemes initiated a trend of structural reduction in flaring over the past two decades.

As figure 4.4 shows, most of the flaring reduction was achieved in the first 15 years of this century, after which flaring volumes plateaued for several reasons. Specifically, the flaring fines in place up to that point were too low and enforcement capacity was insufficient to effectively discourage oil companies from flaring. Most of the flaring reduction before 2015 involved large and easy-to-monetize flare sites, leaving the smaller and trickier sites to be dealt with at a future date. Furthermore, in this forbidding regulatory environment, international oil companies (IOCs) had no incentive to outsource FMR to independent developers with an interest in tackling smaller projects—providing the developers with access to flare sites and data.

Nigeria's flare sites are spread across the Niger Delta, both onshore and offshore, and differ greatly in their size and their proximity to gas pipeline infrastructure. The government estimates that in 2019 flaring was responsible for 22 million tCO_2e emissions or US$385 million in shadow carbon credit value. From an economic standpoint, the amount flared in 2019 equaled 53 million barrels of oil equivalent, two to three LNG trains, and some 3,300 MW of power generation capacity.

In 2016, the government stepped up its efforts, announcing the goal of zero routine flaring by 2020 and launching the development of a market-driven scheme to eliminate flaring, the Nigerian Gas Flare Commercialisation

Programme (NGFCP). The study culminated in the approval of the Flare Gas (Prevention and Pollution) Regulations in July 2018 and the corresponding guidelines in December 2019. These regulations

- Introduced a new and more onerous penalty regime for flaring based on the "polluter pays" principle, with fines of US$0.50 and US$2.00 per thousand standard cubic feet (mscf) of gas flared for producers or, respectively, less than 10,000 barrels of oil per day and 10,000 or more barrels;

- Mandated, by means of the corresponding guidelines, that producers report flaring data, including mandatory installation of meters and penalties for failure to produce accurate flare data or provide access to flare sites; and

- Established the legal basis for the implementation of the market-driven scheme, the NGFCP, giving the government the power to issue permits to access flare gas sites and take associated gas, and to grant such permits via competitive bids.

Program description

The goal of the NGFCP is to eliminate routine flaring via market-driven transactions and commercial structures that attract credible third-party investors deploying proven technologies to monetize associated gas. Such structures should ensure bankability for investors and lenders, preserve the integrity and safety of upstream operations, and benefit Niger Delta communities. The program was launched by the Ministry of Petroleum Resources and set up with the support of donors including the Global Gas Flaring Reduction Partnership and the US Agency for International Development. Overall responsibility for the program was subsequently transferred to DPR. Gas Strategies, a British gas consulting firm interviewed for this case study, was one of the advisers on structuring the program.

NGFCP was established after a three-year planning period, during which the Ministry of Petroleum Resources and the Department of Petroleum Resources were assisted by the World Bank and the Global Gas Flaring Reduction Partnership. The planning phase included interactions with IOCs, midstream companies, financial institutions, and government stakeholders. Four priorities guided the planning phase: (1) introducing meaningful flare payments that would "bite" but at the same not completely destroy the economics of oil production at the affected fields, (2) allowing access to flare sites to third parties through a transparent process, (3) allowing for market dynamics to price associated gas through competitive processes, and (4) implementing a robust data management system.

Under the program, DPR allocates to eligible FMR developers the rights to implement FMR projects and monetize associated gas at specific flaring sites (one or multiple) through a competitive bidding process. The program is agnostic as to the FMR solutions proposed by bidders (gas exports are also allowed) as long as bidders meet technical and financial eligibility criteria and their bids are the most competitive. Examples of solutions mentioned by NGFCP include substitution of higher-cost fuels, such as diesel used in captive power generation; production of liquefied petroleum gas, adding to the existing Nigerian supply or substituting imports; provision of natural gas to new industries, especially in areas without access to pipelines, where compressed and liquefied natural gas could make sense; transformation of gas into electricity, petrochemicals, or fertilizers; and use of gas for small-scale power generation to supply local communities not connected to the grid (Nigeria, Department of Petroleum Resources 2019).

The program is structured in four phases (figure 4.5): (1) initial screening of FMR developers for compliance with minimum technical and financial qualifications, (2) detailed proposals submitted by bidders based on flare site data and information collected by DPR, (3) evaluation and selection of winning bids based on a combination of technical and financial criteria, and (4) signing of agreements and award (for the agreed-on fee) of permits to access and monetize flare gas. To ensure transparency and accountability, DPR set up a web portal for information sharing, submission, and evaluation of bid documentation. Payment of fees and bonds is required at various stages to weed out non-serious bidders.

The qualification phase concluded in July 2019, with 203 developers passing this initial screening. Applicants had to submit standard corporate information and information regarding their technical capabilities and financial capacity. The screening criteria were relatively loose, in consideration of the diversity of flaring situations and potential technical solutions. Large financial capacity, for instance, may not be needed for FMR investments at small flares or with the potential to rent, rather than purchase, equipment. Oil producers were also allowed to apply for the flare sites in their fields, but only through a midstream corporate entity incorporated in Nigeria and under several conditions.

In the bid phase, the first round of which concluded in June 2020 after COVID-19-related delays, qualifying developers submitted bids for rights to implement FMR projects at one or more flare sites. Bidders obtained access to an online data room compiled by DPR on the basis of information provided by oil producers and a request for proposal package explaining the process and requirements for bids. DPR identified a total of 178 flare sites but included only 48 in the first auction, for a total of about 300 mmscf/d. The sites included in the first auction met the following criteria: (1) investment-grade available data on

FIGURE 4.5

Nigerian Gas Flare Commercialisation Programme implementation phases

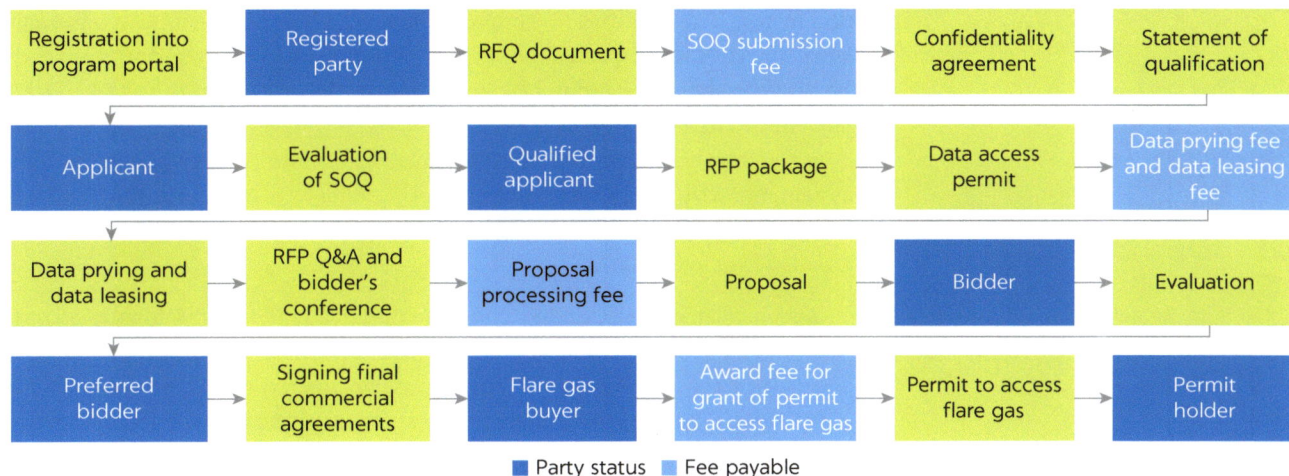

Source: Nigeria Department of Petroleum Resources (https://ngfcp.dpr.gov.ng/media/1145/dpr-presentation-overview-of-the-ngfcp-rev-4.pptx).
Note: RFP = request for proposals; RFQ = request for qualifications, SOQ = statement of qualifications.

these sites, (2) a minimum flare size of 1 mmscf/d sustainable for seven years, and (3) absence of any legal disputes involving the sites. The data room included information on flare sites, flare volumes (including forecasts and depletion rate estimates by the oil producers), associated infrastructure, and geographic and socioeconomic data. Each oil producer was required to provide the annual amounts of flare gas that it expects to have available for a minimum of 15 years, or the expected life of the oil field. Bidders had the ability to request site visits in coordination with DPR (it is unknown to what extent bidders took advantage of this opportunity). Additional information was provided to all bidders through formal Q&A lists.

Proposals were submitted in two parts:

- *Envelope 1*. Technical and commercial proposal, with the technical details, business case, and all other underpinning economic assumptions of the proposed project(s).

- *Envelope 2*. Financial proposal, including (1) the price the bidder committed to pay for associated gas in US$/mscf, and (2) the gas volume the bidder offered to contract, split between a guaranteed volume for each contract year under a take-or-pay agreement (minimum 70 percent of the total bid quantity) and a nonguaranteed volume. Bidders were free to bid for any volume below the flare gas forecast quantity, and the volumes could vary for each contract year.

In the bid evaluation phase, an evaluation committee appointed by DPR selected winners on the basis of the two envelopes. Only proposals that passed the technical and commercial evaluation had their financial proposals opened and advanced to the final evaluation. Preferred bidders were selected on the maximum aggregated value of the flare sites for which they bid, determined as the net present value of the take-or-pay quantity multiplied by the sum of the flare gas price and a shadow emission credit price. The latter was the same for all bidders, and therefore its inclusion in the evaluation formula implicitly favored bids for larger volumes of associated gas—consistent with the environmental objectives of the program. Under the NGFCP, the government retains all emissions reduction credits, with the possibility to monetize them in the future. The unit value of the emissions credit for each contract year was set by DPR (in US$/mscf terms). The proposals were ranked in order of decreasing total net present value. The DPR was expected to complete the selection process and allocation of the 48 flare sites to winning bidders by end of June 2021, but the COVID-19 pandemic halted the process.

In the final phase, preferred bidders will enter into a series of commercial agreements with the government and oil producers, after which they will be awarded permits to access flare gas. These agreements are the following (see schematics in figure 4.6):

- *Gas supply agreement*, under which the preferred bidder buys associated gas from the government at the price stipulated in the bid. The agreement also specifies the take-or-pay obligation, performance bonds, and penalties. Invoicing is based on a revenue meter at the metering point. This agreement alone does not provide security of flare gas supply to the preferred bidder, because the oil producer is not a party to the agreement. The oil company operates the equipment that connects the flare gas tie-in to the FMR project.

FIGURE 4.6

Schematics of Nigerian Gas Flare Commercialisation Programme commercial agreements

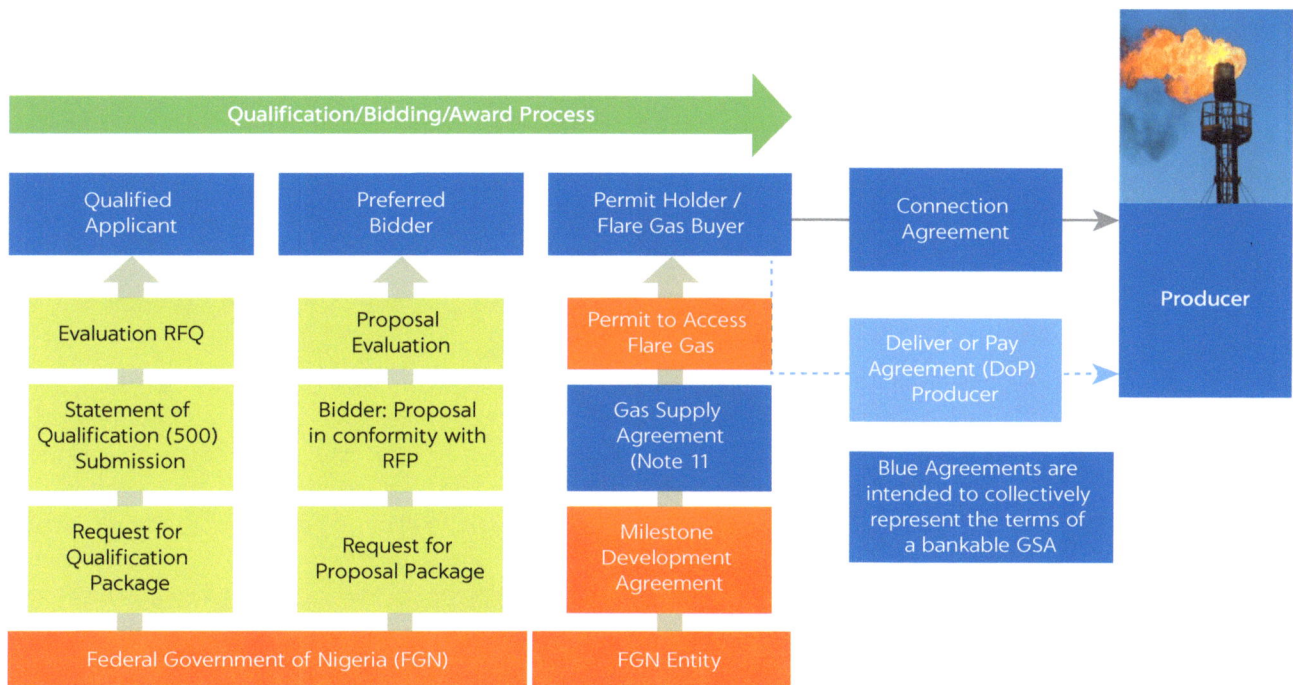

Source: Nigeria, Ministry of Petroleum Resources 2019.
Note: The intent is to see whether a limited "first loss" provision (deliver or pay) for contracted nonguaranteed flare gas can be developed. Agreements in blue are intended to collectively represent the terms of a bankable gas supply agreement. RFP = request for proposals; RFQ = request for qualifications.

- *Milestone development agreement,* under which the preferred bidder provides a financial guarantee to the government to underpin its commitment to implementing project milestones.
- *Connection agreement* between the preferred bidder and the oil producer, stipulating the arrangements pertaining to flare gas delivery infrastructure. The agreement authorizes the preferred bidder to engineer, procure, and construct the gas connection infrastructure at its own expense and under criteria acceptable to the oil producer (for example, approved vendor list). Ownership of the infrastructure is then handed over to the oil producer, which will be responsible for operations and maintenance and will receive a handling fee in compensation. Handling fees will be in accordance with a schedule provided by DPR and based on maximum capacity throughput.

- An optional *deliver-or-pay agreement* between the preferred bidder and the oil producer, with the latter providing a guarantee of gas quantity meeting specification limits set by the NGFCP.

Takeaways

NGFCP was designed as a competitive mechanism to attract developer capital and expertise to the implementation of FMR projects at small and complex flare sites. Even with meaningful flare fines in place, IOCs may not have sufficient economic incentives to directly implement small FMR projects that are perceived as noncore and whose profit potential is trivial compared to that of the core oil-producing operations. Difficulties in enforcing flare payments further disincentivize IOCs from taking the initiative to address flaring. In this context, the Nigerian government, after careful planning, opted for a system that relies on third-party developers to implement FMR projects under a competitive selection process and with the obligation for IOCs to allow such developers access to flare sites.

Because no FMR project has yet been implemented under NGFCP, it is impossible to draw conclusions about the success of the program.

CRUSOE ENERGY SYSTEMS

Background

For the growing number of technology companies relying on energy-intensive computing, availability of inexpensive power is key to profitability. Computing applications such as Bitcoin mining, running of artificial intelligence algorithms, and image rendering sometimes require vast amounts of power. A study by the University of Cambridge estimates that Bitcoin "mining"[11] uses more electricity annually than all of Argentina, the 30th-largest country by energy use (Blandin et al. 2020; Criddle 2021). Bitcoin prices on the one on hand and mining costs (primarily electricity and computing equipment) on the other determine the profitability of a cryptocurrency miner.

In recent years, power-hungry Bitcoin miners and other computing-intensive businesses have set up or shifted operations to regions where electricity, usually generated from hydro or coal, is cheap, often regardless of carbon emissions. Sixty-two percent of miners surveyed globally for a study by the University of Cambridge reported using hydroelectricity; coal came in second (38 percent of respondents), followed by natural gas (36 percent). The breakdown in the amount of energy used by source is a hotly debated topic that is outside the scope of this case study. By Cambridge's estimates, only 39 percent of Bitcoin mining's energy consumption is from renewable sources—a figure that would reflect the high concentration of Bitcoin mining operations in coal-reliant regions of China and Mongolia (Blandin et al. 2020). The same study, however, also notes the argument that Chinese Bitcoin miners switch to hydroelectric energy in the rainy season (when China has an oversupply of it), which would increase the overall reliance on renewables.

In response to this trend, a few start-ups have emerged in the United States that use associated gas to power cloud computing operations such as Bitcoin mining. Among these are Crusoe Energy Systems ("Crusoe"), EZ Chain, and Wesco. These start-ups differ in their business models but all have the same mission: to reduce the reliance of computing-intensive businesses on electricity from the grid, which in the United States is 19 percent coal-based,[12] providing cheap power from gas that would otherwise be flared. This phenomenon is not unique to the United States: the Russian state-owned oil company Gazprom is mining Bitcoin with associated gas in West Siberia (Davis Szymczak 2021).

This FMR solution is gaining ground among shale oil producers in the United States, in response to several trends, including the following:

- *A challenging oil price environment* that forces producers to explore any revenue opportunities, including the sale of associated gas.

- *A tightening regulatory environment.* In the United States, flaring regulations vary by state. Each state has provided different incentives or mechanisms to minimize flaring, usually with allowances for extraordinary circumstances. However, most states are trending toward implementing regulations that could have a negative financial impact on companies that flare associated gas regularly.[13] Furthermore, the recently announced climate change policy of the US government specifically targets the reduction of gas flaring and methane emissions in oil and gas operations in forthcoming federal regulations.[14] A flare mitigation tax credit has partially passed in the North Dakota state legislature to incentivize the use of flare mitigation technologies such as digital flare mitigation.[15]

- *Operational and infrastructure challenges* include the absence of midstream infrastructure near producing wells, delayed pipeline arrival, pipeline capacity challenges, and extended gas plant (if any) downtime.

- *Increasing investor focus on the environmental standards of US oil-producing companies.* For instance, Blackrock, one of the largest asset managers globally, stated that, in order to track a net zero goal by 2050, a "near elimination of flaring, with government policies and industry commitment, must occur by 2025" and that the public and private sectors need to work together to deploy existing and emerging flaring reduction technologies (Blackrock 2021). In the case of Texas, investors managing over US$2 trillion in assets are calling on the state's regulators to ban the routine flaring of gas from shale fields. See chapter 2 for more information on investor sentiment (IEEFA 2020).

Crusoe energy systems' digital flare mitigation

Crusoe is a Colorado-based, Silicon Valley–funded start-up that converts associated gas into electricity for energy-intensive computing at the well site, a solution it calls "Digital Flare Mitigation" (DFM).[16] Crusoe was cofounded by energy industry professional Cully Cavness and tech entrepreneur Chase Lochmiller, and has been operational since 2018. It raised US$150 million in equity capital from venture capital funds including the Founders Fund, Bain Capital Ventures, and Valor Equity Partners. Crusoe is currently running 40 DFM systems in three oil-producing areas in the United States—the Powder River Basin (Wyoming), Williston Basin (Montana and North Dakota), and the DJ Basin (Colorado and Wyoming). Crusoe's clients have included Equinor, the Norwegian national oil company with operations in the Bakken; Kraken Oil & Gas, the largest producer of Bakken oil in Montana; EnerPlus, a Canadian company that holds operating acreage in the Bakken and a nonoperating position in the Marcellus field; Devon Energy, one of the largest publicly traded producers in the Bakken; and other private and public operators in the Denver Julesburg Basin, Powder River Basin, and Williston Basin. One large operator, for example, has deployed 18 DFM modules across six oil and gas production sites in eastern Montana and one site in North Dakota. Most of the operator's wells sell gas into traditional pipelines for processing at gas plants. However, a portion of the company's acreage is in areas with limited or unavailable pipeline capacity, which is where the operator has deployed Crusoe's DFM solution.

Crusoe's DFM systems are designed for turnkey deployment and mobilization in the form of modular and scalable equipment, which allows Crusoe to mitigate flaring at almost any scale from tens of thousands to millions of cubic feet per day. DFM systems are composed of rich-burn power generation equipment (reciprocating engines or turbines), data centers built in containers modified for use and installation in oilfields, and specialized computers connected to the internet via satellite (photo 4.2).

The equipment is modular and can be easily scaled up or down as the size of the flare varies over time (the average lifetime of a DFM project is one to three years). Installation and start-up of equipment take as little as one to two weeks (table 4.1).

The minimum size for Crusoe to deploy is generally approximately 300 mscf/d to support one 2 MW DFM module. For operators without access to utility power at the well, some of the electricity generated by Crusoe can also be used for on-site needs. Master service agreements with the operators ensure Crusoe's compliance with safety procedures, including third-party verification (only one minor recordable safety incident has occurred in the company's history). Crusoe's operations and business development team is assembled from established upstream, midstream, and service companies.

Crusoe believes the opportunity for DFM in the United States is significant because of the volumes flared in the Bakken and the Permian Basin. Crusoe is currently capturing approximately 6 mmscf/d across operations in the Bakken.

PHOTO 4.2
Example of Crusoe's Digital Flare Mitigation system

Source: © Crusoe Energy Systems. Used with the permission of Crusoe Energy Systems. Further permission required for reuse.

TABLE 4.1 **Deployment of Crusoe's Digital Flare Mitigation solution and estimated timeline**

STEP 1: CONNECTION POINT	STEP 2: GENERATOR	STEP 3: COMPUTING MODULE	STEP 4: START-UP
Operator provides simple manifold and valve to existing gas line. Typically, manifold is installed directly onto line leading to flare.	Crusoe provides generator system, delivered on portable trailer or skid.	Computing modules delivered by truck. Satellite antennae installed and aligned after delivery.	Computers connected to generator. Generator start-up. If any gas occurs above and beyond the capacity of Crusoe's DFM system, the flare is still in place to safely combust that gas.
1–2 days	1 day	1–5 days	1 day

Source: Interview with Crusoe Energy Systems.
Note: DFM = digital flare mitigation.

An estimated 250 mmscf/d to 600 mmscf/d is flared in the Bakken, which could support the deployment of more than 1,500 DFM systems. Flaring in the Permian Basin is estimated at greater than 600 mmscf/d.

With regard to its business model, Crusoe (1) purchases, installs, and operates the DFM modules at no cost to the oil producer; (2) remunerates the producer by purchasing associated gas; and (3) generates revenues from either Bitcoin mining or operating high-performance computing (HPC) data centers for third-party clients.[17] The power generation units represent typically one-third of capex, the data center shell (customized container) another 10–20 percent, and the remainder is for computer systems. The price paid for associated gas is generally lower than pipeline prices but sufficient to incentivize operators not to flare. When mining Bitcoins, Crusoe mines the cryptocurrency for itself, which exposes equity returns to the variability in Bitcoin prices.[18] When providing cloud computing services to third parties, Crusoe has several business models, including fixed rack space and power capacity fees or variable charges per unit of time based on the type of computing hardware used by the customer.

Crusoe's solution is advantageous for oil producers because it (1) ensures compliance with flaring regulations, (2) creates an additional revenue stream from associated gas, and (3) demonstrates the producers' commitment to reducing emissions. The concessionaire (via royalties) and the states (via severance taxes) also benefit financially from Crusoe's solution. For instance, Crusoe has been servicing a large public exploration and production operator drilling horizontal Bakken oil and gas wells in McKenzie County, North Dakota. The operator has access to a gas-gathering pipeline system that lacks sufficient capacity to take all of the gas from the five wells on the site. The volume of excess gas is variable and dependent on the gas production from all of the other well sites on the gas-gathering system. In this situation, the regulator would have allowed the operator to flare volumes. Alternative options, such as gas-to-liquids or compressed natural gas, were uneconomic for the operator, as was a pipeline capacity expansion for the gas-gathering company. Crusoe installed three DFM modules consuming approximately 1 mmscf/d of otherwise flared gas for at least one year. The pipeline company was willing to give the operator a release from its gas dedication agreement for 1 mmscf/d for at least one year to allow Crusoe to use its DFM system to purchase gas and consume it (as opposed to business as usual and flaring the gas). Crusoe takes possession of the gas (rich, at 1,500 Btu) directly from the wellhead after traveling through a three-phase separator. Additionally, the operator does not have access to utility power at the well site but is able to use a small subset of the excess gas produced to generate power for on-site uses.

A DFM project's capital structure consists of Crusoe equity and debt provided by specialized equipment lenders. Power generators are financed with amortizing loans from OEMs or third-party finance providers, requiring only a small down payment (20 percent) up front from Crusoe. The financing market for cloud computing equipment is much less developed, especially for Bitcoin mining operations, whose profitability is directly linked to volatile Bitcoin prices and fluctuating network hashrate.[19] Crusoe managed to secure a US$40 million project financing facility from Upper90, secured against computing equipment and with an accelerated payback period.[20] As the Bitcoin market becomes more liquid and hedging solutions may emerge, banks may also enter the financing market for Bitcoin mining equipment. Currently, Bitcoin hedging solutions are very limited, but Crusoe believes its low-cost power provides resiliency in a scenario of worsening mining economics.

Crusoe's DFM solution results in both direct and indirect greenhouse gas emissions reductions. Crusoe estimates that its DFM system provides a 98 percent reduction in methane emissions and 41 percent CO_2e reduction when compared to a pit flare (considering that methane is not completely combusted in well site flares), and 25 percent CO_2e reduction compared to an open flare.[21] When accounting for the frequency of unlit or malfunctioning flares, the company estimates that CO_2e reductions reach up to 63 percent when compared to status quo flaring.[22] In addition, Crusoe's modular data centers reduce power demand from the grid, which in the United States has an average Grid Emissions Factor of 0.6581 tCO_2e for each megawatt-hour of electricity generated. Crusoe's generators have catalytic converters and emissions control kits, with methane combustion efficiency measured at 99.89 percent.

Takeaways

Crusoe provides a capital-light flare mitigation approach, taking exposure to the fluctuations in Bitcoin mining economics. Crusoe's DFM solution aims to provide a positive financial and environmental outcome for all stakeholders: operators, environmental groups, regulators, and royalty owners. Crusoe bears all of the risk of Bitcoin mining economics but believes that its access to cheap electricity from flare capture hedges some of the Bitcoin mining risk. Specifically, if Bitcoin prices drop, mining economics will degrade and miners with higher-cost electricity or less efficient computers will likely be forced to drop off the network. This situation would decrease the network hashrate and will increase Crusoe's share of Bitcoin mined, even at a lower Bitcoin price.

DFM is an opportunity to meet the growing demand for energy-intensive computation with the abundant energy from natural gas that would otherwise be flared. Although there is effectively no constraint on the ability to monetize gas or power through energy-intensive computing (because of its modularity and scalability), the challenge is in the ability to source the quantum of computers to meet the demand for DFM. Currently, supply chain issues—including competition for chips from large corporates such as Apple—limit Bitcoin miners' access to state-of-the-art computers (Roberts 2020). Additionally, given the energy density of natural gas, it can take hundreds or even thousands of computers to consume the electricity generated from a shale well flare gas stream—adding to the capex needed for operation.

At the same time, not all DFM solutions apply to all flare sites. Initial production decline curves from shale wells are challenging to match. Crusoe prefers to be a baseload consumer of stranded gas rather than deploy and remove equipment when the initial production volumes decline rapidly after a month or two. In addition, the declining gas supply of some flare sites is a challenge, especially for regular data center services that need stable power. Regular data center services also have high bandwidth requirements, which are often unattainable in remote locations like oil sites. Conversely, Bitcoin mining offers more flexibility because it requires limited networking capacities and is interruptible.

NOTES

1. A deliver-or-pay agreement binds the oil company to deliver a guaranteed volume of associated gas or pay penalties.

2. Currency risk is particularly evident in FMR projects that sell the end-product in local currency but—in most cases—purchase associated gas in US dollars.

3. For more information, see the Hoerbiger website (https://www.hoerbiger.com/en-3/pages/549).

4. Revenue split provided by Hoerbiger executives interviewed. Unless otherwise noted, other data in this case study also come from interviews.

5. It must be noted, however, that flaring fines are not in place in Ecuador, despite the Zero Routine Flaring 2030 commitment.

6. In some cases, the projects involve the recovery and overhaul of stranded/unused assets of the customer instead of new equipment reducing sunken cost.

7. The US$0.20/kWh cost includes fuel and operation and maintenance costs; US$0.26/kWh represents the cost of incremental diesel generation capacity, including capex in addition to the aforementioned costs.

8. The actual figure is confidential.

9. Unless otherwise stated, the information provided in this case study is sourced from interviews with Galileo's management, a presentation provided by management, and Galileo's website.

10. The lower the atmospheric temperature, the longer the container storage period.

11. The process of creating new Bitcoins by solving computational puzzles.

12. From the US Energy Information Agency's "Energy Explained: Electricity in the United States" web page (https://www.eia.gov/energyexplained/electricity/electricity-in-the-us.php).

13. For instance, North Dakota has increasing gas capture requirements over time that could force operators to curtail oil production if they do not meet the gas capture percentage (https://www.dmr.nd.gov/oilgas/GuidancePolicyNorthDakotaIndustrial Commissionorder24665.pdf). New Mexico has recently announced a new rule that requires operators to reach 98 percent gas capture by 2026 (Executive Order 2019-003, Energy, Minerals and Natural Resources Department of New Mexico).

14. Executive Order on Protecting Public Health and the Environment and Restoring Science to Tackle the Climate Crisis, January 2021 (https://www.whitehouse.gov/briefing-room/presidential-actions/2021/01/20/executive-order-protecting-public-health-and-environment-and-restoring-science-to-tackle-climate-crisis/).

15. North Dakota Bill SB 2328, signed into law by the governor on April 13, 2021 (https://www.legis.nd.gov/assembly/67-2021/bill-actions/ba2328.html and https://www.legis.nd.gov/assembly/67-2021/documents/21-0935-05000.pdf).

16. Unless otherwise stated, the information in this case study was provided by Crusoe's management.

17. As of May 2021, only one of Crusoe's sites has an HPC data center in addition to Bitcoin mining. The company expects to grow the HPC business significantly in 2021. Not all sites are suitable to HPC, because the smooth running of operations requires redundant power (or grid backup) and enhanced networking.

18. Bitcoin miners used advanced computers to solve complex math problems, in competition with other miners. The computer that solves the problem adds a block to the blockchain, a secure ledger that serves as a public record of transactions. The owner of the winning computer obtains a block reward of 6.25 Bitcoins. The process is repeated every 10 minutes or so (Hackett 2016).

19. Hashrate is a measure of the level of competition among Bitcoin miners.

20. Information derived from https://alejandrocremades.com/chase-lochmiller/ and https://www.upper90.io.

21. Information from a Crusoe presentation with analysis prepared by independent third-party emissions consultant for illustrative 1,800 mscf/d project with 1,500 mmBtu/mcf gas composition using US Environmental Protection Agency and North Dakota Department of Environmental Quality methodologies.

22. Based on EDF (2022) and a Crusoe presentation with analysis prepared by independent third-party emissions consultant for an illustrative 1,800 mscf/d project with 1,500 mmBtu/mcf gas.

REFERENCES

Blackrock. 2021. "Climate Risk and the Transition to a Low-Carbon Economy." *Investment Stewardship Commentary*, February. https://www.blackrock.com/corporate/literature /publication/blk-commentary-climate-risk-and-energy-transition.pdf.

Blandin, Apolline, Gina Pieters, Yue Wu, Thomas Eisermann, Anton Dek, Sean Taylor, and Damaris Njoki. 2020. "3rd Global Cryptoasset Benchmarking Study." Cambridge Centre for Alternative Finance, University of Cambridge, UK https://www.jbs.cam.ac.uk/wp -content/uploads/2021/01/2021-ccaf-3rd-global-cryptoasset-benchmarking-study.pdf.

Colectivo (Colectivo eliminen los mecheros que encendemos la vida). 2020. "Informe: Mecheros en Ecuador." Colectivo, Quito. https://redamazonica.org/wp-content/uploads /Informe-MECHEROS-EN-ECUADOR.pdf.

Criddle, Cristina. 2021. "Bitcoin Consumes 'More Electricity Than Argentina.'" BBC News, February 10. https://www.bbc.com/news/technology-56012952.

Davis Szymczak, Pat. 2021. "Gazprom Neft Mines Bitcoin as an Alternative to Flaring Unwanted Gas." *Journal of Petroleum Technology*, January 13. https://jpt.spe.org/gazprom -neft-mines-bitcoin-as-an-alternative-to-flaring-unwanted-gas.

EDF (Environmental Defense Fund). 2022. "Methodology for EDF's Permian Methane Analysis Project (PermianMAP)." Data and Collection Analysis, EDF, New York. https://www.edf .org/sites/default/files/documents/PermianMapMethodology_1.pdf.

El Comercio. 2021. "Corte de Ecuador acepta demanda de niñas para eliminar mecheros en la Amazonía." January 26. https://www.elcomercio.com/actualidad/ecuador/corte-demanda -ninas-mecheros-amazonia.html.

Hackett, Robert. 2016. "Wait, What Is Blockchain?" *Fortune*, May 23. https://fortune .com/2016/05/23/blockchain-definition/.

IEEFA (Institute for Energy Economics and Financial Analysis). 2020. "Leading Investment Firms Push Texas Regulators to Crack Down on Natural Gas Flaring." IEEFA, September 8. https://ieefa.org/leading-investment-firms-push-texas-regulators-to-crack-down-on -natural-gas-flaring/.

Nigeria, Department of Petroleum Resources. 2019. "The Nigerian Gas Flare Commercialisation Programme." Programme Information Memorandum, Federal Government of Nigeria, Abuja. https://ngfcp.dpr.gov.ng/media/1134/ngfcp-pim-rev1.pdf.

Roberts, Jeff John. 2020. "The American Heartland Needs Jobs. Could Bitcoin Mining Become Its Next Savior?" *Fortune*, December 12. https://fortune.com/2020/12/12 /bitcoin-jobs-cryptocurrency-mining-hiring-core-scientific/.

5 Practical Considerations for Implementing Flaring and Methane Reduction Projects

FINANCIAL ATTRACTIVENESS OF FLARING AND METHANE REDUCTION INVESTMENTS

This report aims to create awareness of the business case for reducing gas flaring and methane emissions, the financial attractiveness of flaring and methane reduction (FMR) projects, and the barriers that project developers need to overcome. The study analyzed the financial attractiveness of six FMR solutions at flare volumes of 1 million standard cubic feet per day (mmscf/d), 5 mmscf/d, and 10 mmscf/d: (1) gas-to-power (with electricity used externally), (2) gas-to-power (with electricity used on-site by the oil operator), (3) gas delivery to an existing pipeline network, (4) gas delivery to an existing gas processing plant, (5) compressed natural gas, and (6) small-scale liquefied natural gas. Other solutions exist but were not modeled because of their niche nature (for example, Crusoe Energy's digital flare mitigation, presented in chapter 4). Financial returns (internal rates of return [IRRs] and net present values) were modeled on the basis of indicative assumptions derived from project experience and feedback from industry experts. Sensitivities were conducted to reflect the variability of assumptions in real-life projects.

The analysis points to a potentially attractive financial opportunity for independent developers to invest in FMR projects tackling flares in the study range. About 2,358 flare sites fall in the 1–10 mmscf/d range, representing 53.8 percent of global flare volumes, offering a very significant emissions reduction opportunity. Oil companies are unlikely to divert capital and engineering resources to small noncore projects that, from a profitability standpoint, are trivial compared to their core activities. Although far from straightforward, FMR projects of this size are less complex than projects involving large and mega flares, which require large infrastructure investment (for instance, in gas or electricity transmission infrastructure), government planning, and large capital injections. At the other end of the spectrum, FMR projects at flares smaller than 1 mmscf/d are unlikely to be economically viable, unless clustered in larger projects.

Financial modeling presented in chapter 3 shows that FMR projects at flare volumes between 5 mmscf/d and 10 mmscf/day are potentially attractive investment opportunities. For a typical 5 mmscf/d flare, the unlevered and pretax IRR ranges from a barely acceptable 7 percent for a gas delivery to gas processing solution, to an attractive 20 percent for a small-scale liquefied natural gas solution. A single-digit IRR is below the minimum return threshold of 10 percent suggested by industry participants consulted for this study and would result in a negative net present value. For a 10 mmscf/d flare, all FMR solutions modeled in chapter 3 would produce positive net present values and double-digit IRRs, ranging from 12 percent for a gas delivery to gas processing solution to 24 percent for a small-scale liquefied natural gas solution.

On a standalone basis, 1 mmscf/d flares do not offer attractive financial returns, but they can be clustered to reach an aggregate project size closer to 5–10 mmscf/d. None of the FMR solutions analyzed is financially attractive at flare sites of 1 mmscf/d, with the exception perhaps of power generation for on-site use. The relative advantage of this solution is the ability to charge electricity prices above grid levels, displacing more expensive diesel- or oil-fueled generation. Even this solution, however, would yield only a 7 percent IRR under the model's assumptions. All other solutions are too intensive in terms of capital expenditures (capex) to generate attractive returns at 1 mmscf/d flares. In practice, FMR developers will want to cluster small flares to reach aggregate volumes more in line with the 5 mmscf/d and 10 mmscf/d scenarios. These size considerations apply to the six FMR solutions modeled in chapter 3. Other niche FMR solutions may be financially attractive at smaller flare sites. Crusoe Energy's digital flare mitigation, for instance, is viable at flare sites as small as 300,000 scf/d.

FMR projects at 5 mmscf/d–10 mmscf/d flare sites (unique flares or clusters) involve a capital investment in the range of US$7 million to US$59 million, depending on the FMR solution adopted and according to the model's assumptions. In the 5 mmscf/d scenario, capex would range indicatively from US$7 million (gas delivery to existing pipeline) to US$31 million (gas-to-power for external use, with 17 megawatts capacity installed). In the 10 mmscf/d scenario, capex would range indicatively from US$13 million (gas delivery to existing pipeline) to US$59 million (gas-to-power for external use, with 38 megawatts capacity installed). Please refer to chapter 3 for full details.

Projects of these sizes can be financed primarily with equity provided by the developer and co-investors. This is important because many FMR projects do not meet the requirements of traditional nonrecourse project finance, at least in the development phase. Many FMR projects do not meet contractual project finance criteria—for instance, they may not have feedstock supply agreements in place. In addition, tight development timelines may limit the window of opportunity to contractually arrange a debt package. When a project meets the requirement for debt finance, the magnitude and terms of such debt may not be as attractive as in more established project finance sectors. The Thermo Mechero Morro gas-to-power project discussed in chapter 4, for instance, managed to raise debt but only to the tune of 55 percent of the capital structure.

In the operational phase, when development risk is no longer present, FMR projects may be easier to leverage—especially if de-risking tools are available. FMR developers and investors interviewed for this study indicated use of such tools as a potential avenue to recover some of the equity investment before project expiration (see the Hoerbiger case study in chapter 4). However, not all risks disappear in the operational phase: for instance, FMR projects remain exposed to off-taker risk. De-risking tools such as partial credit guarantees or insurance for breach of contract could enhance the bankability of FMR projects, allowing developers to take money off the table and reallocate capital to new FMR projects.

RISKS OF FMR INVESTMENTS AND MITIGATING FACTORS

Despite offering attractive returns on paper, FMR projects are subject to a variety of risks that must be seriously evaluated and can, to an extent, be mitigated. These risks fall primarily into five categories:

1. *Associated gas supply risk.* Often, the quantity and quality of associated gas are not fully known up front, particularly when the oil company has kept a poor record of its flaring activities, perhaps because of inadequate regulation and enforcement. The future decline rate of a flare is also a significant unknown, and forecasts in that respect are far from reliable. Several strategies can be considered to mitigate this risk:

   – Project capex should be sized on the basis of a conservative estimate of associated gas volume during the project duration, to avoid being left with ample spare capacity. This sizing could take the form of an estimated average volume flared assuming a conservative decline rate. Alternatively, capex should be tuned to the estimated volume after two or three years of decline rather than at inception.

   – Clustering several flares will allow developers to hedge their bets and make up any shortfall from a specific flare with gas from another one.

   – Using modular and movable equipment will allow developers to increase or decrease capacity efficiently and shift equipment among different flares at the target site.

   – Where possible, nonassociated gas should back up associated gas to mitigate supply risk and extend the economic life of the projects, allowing for investment to be amortized over a reasonable term and enhancing the financial attractiveness of FMR projects.

   – Factor in variations in associated gas composition in project design criteria. Often the trade-off for higher flexibility in this respect is accepting a lower conversion yield or a higher capex in order to minimize potentially serious constraints over the project life cycle.

   – Developers can attempt to sign a deliver-or-pay agreement binding the oil company to deliver a guaranteed volume of associated gas or pay penalties, although oil companies are likely to resist this arrangement.

2. *End-product price risk.* The financial returns of FMR projects are significantly affected by changes in end-product prices (electricity or gas), as demonstrated in chapter 3. This risk is more prevalent in projects selling the end product externally and is muted, or absent, in gas-to-power projects selling electricity for the oil company's own use; because a reliable power supply is critical to field operations, oil companies are usually willing to lock in an electricity price for the duration of the FMR project. Conversely, external off-takers are in many cases not willing to sign long-term contracts. The strategies to mitigate this risk include the following:

   – Negotiating with the oil company a tolling fee remuneration structure that shifts (at least partially) onto the company the risk of end-product price volatility;

   – Evaluating in the project design phase the risk that new capacity from competing fuels (including renewable energy) may negatively affect market prices in the future;

- Installing modular and movable equipment to avoid being left with stranded or underused assets;
- Signing offtake agreements for periods shorter than the target project duration, to eliminate volatility at least partially over the project life cycle; and
- Building a diversified portfolio of FMR projects with different off-takers or end markets, if feasible.

3. *Off-taker payment risk.* FMR projects are exposed to the risk that the off-taker (for example, power distribution company) may not honor its contractual obligations, perhaps as a result of financial difficulties unrelated to the FMR project itself. This risk is not dissimilar from the risk incurred by many infrastructure projects and may significantly affect the ability to leverage the FMR project with nonrecourse debt. The strategies to mitigate this risk include

- Seeking guarantees from the off-taker parent company or controlling entity, especially if it is the government;
- Seeking guarantees or insurance for breach of contract from commercial or concessional providers, including development banks;
- Evaluating the creditworthiness and reliability of the off-taker through due diligence, including analyzing any history of nonpayments in other projects; and
- Where feasible, building a diversified portfolio of FMR projects with different off-takers.

4. *Project execution risk.* FMR projects include many moving parts and stakeholders. They are often in remote locations with limited infrastructure. If infrastructure needs to be built, the buy-in of local stakeholders may be necessary (for example, for obtaining the necessary rights of way for power transmission lines). Delays can occur during manufacturing or construction, along with inconveniences in other links of the value chain (downstream or upstream), which are out of the project sponsors' control. The only mitigating strategy against these risks is to design watertight engineering, procurement, and contracting and to employ reputable contractors and established original equipment manufacturers. The latter also increasingly act as turnkey FMR project developers. By using and being in control of their own equipment, original equipment manufacturers-developers can exercise greater control over project execution.

5. *Macroeconomic and political risks.* FMR projects face a variety of macroeconomic risks, including those affecting oil companies. A significant drop in oil prices, for instance, may affect production at a given oil field and with it the supply of associated gas. Macroeconomic volatility may also affect prices and demand of the FMR project's end product. Currency volatility could have severe repercussions for some FMR projects: in many countries the price of associated gas is in US dollars, loans are denominated in US dollars or other hard currencies, but off-takers' tariffs and tolling fees are in local currency terms, exposing projects to local currency depreciation risk. Political risks in oil-producing countries include extended operational disruptions, often attributable to social unrest or pipeline damage, nationalization of oil fields or forced renegotiations of terms and conditions, and changes in legislation affecting the oil industry. These risks are not unique to the FMR sector,

and some are simply part of the business for the oil and gas sector. Mitigating strategies include

– Negotiating force majeure clauses in associated gas supply contract and offtake agreements, although the negotiating power of independent FMR developers may be limited;

– Political risk insurance from commercial or concessional providers, including development banks, if available at affordable terms;

– Currency hedging, if available at affordable terms in the project country; and

– Multicountry portfolio diversification, where feasible.

Although this report focuses on a project-level analysis of returns and risks, FMR projects are heavily affected by the national regulatory framework with regard to flaring, venting and methane emissions. Critical regulatory aspects to be investigated in project design phase include the following:

• *Ownership of the associated gas.* In many jurisdictions or contracts, who owns the associated gas is not defined. In most cases, oil companies are authorized to use associated gas for own consumption or reinjection, but rules regarding associated gas monetization may not be as clear-cut.

• *Flare and emissions disclosure rules* and their impact on the accuracy of the data provided by the oil companies to FMR developers.

• *Flare and methane emissions payments and penalties* and their indirect impact on FMR project economics. For instance, an oil company subject to heavy fines for flaring may be prone to supplying associated gas to the FMR developer at very advantageous prices.

• *Rules providing or denying third-party access to oil and gas facilities,* which could affect an FMR developer's ability to establish operations on-site.

• *Tax rules,* including whether an FMR project is treated as an upstream or (more likely) midstream player and the ability to take advantage of accelerated depreciation schemes.

APPENDIX A

Sample of Service Companies that Offer Flaring and Methane Reduction Solutions

TABLE A.1 **Selected companies that offer flaring and methane reduction solutions**

COMPANY	PROJECT EXAMPLES
Generon	No projects given, but the company seeks to offer economical alternatives to flaring, including flare gas power generation, flare gas reinjection, flare gas used as feedstock in petrochemicals, and compressed natural gas.
Capterio	No project examples given, but the company aims to partner with energy companies to deliver flare monetization projects and to bring together assets, methods, and financing to do so. Capterio delivers on-the-ground abatement projects for clients. Projects include reinjection, raw gas sent to the nearest export pipeline, power generation, and the recovery of liquids or conversion of gas to liquids for sale.
Baker Hughes	No project examples given, but the company has an entire business line dedicated to methane monitoring and management. It advertises technology to reduce venting, flaring, and fugitive emissions in the oil and gas industry. However, upon further inspection, its offering appears to reduce flaring by just making it more efficient, rather than eliminating it, but is still possibly worth investigating.
Pioneer Energy	Provider of mobile flare gas capture solutions and modular gas processing plants. Pioneer Energy's flare gas capture and processing systems turn raw associated and nonassociated gas and oil tank vapors from waste streams into resources. Systems are skid-mounted, modular, autonomous units that are remotely monitored and controlled, which enables flexibility in equipment deployment and superior uptime while minimizing required capex and opex.
Alphabet Energy	Alphabet Energy's Power Generating Combustor (PGC™) for oil and gas flares has seen strong commercial traction in helping oil and gas operators mitigate permitting risk and meet remote power needs. The product helps expedite or eliminate permitting processes by transforming gas flares or combustors into power generators.
Caterpillar	In 2012, three engineers with decades of experience in the energy industry founded GTUIT with the goal of reducing natural gas flaring in North America and around the world. GTUIT's mobile, modular gas capture and natural gas extraction units are about the size of a semitrailer and easily connect to an engine or generator set. They significantly decrease the volume of flared gas at the wellhead and reduce the volume of volatile organic compounds released into the atmosphere. The units also remove valuable NGLs and produce a dry, consistent, high-Btu gas. Energy companies can conserve and sell the NGLs on the market for later use, and they can use the conditioned gas as free fuel to power on-site gas or dual-fuel engines and generator sets.
SoEnergy International	SoEnergy has a flare gas recovery system that converts flared gas into fuel for upstream operations. SoEnergy's flare gas recovery systems are customized from beginning to end. Treatment and conditioning systems transform even the toughest flare gas compositions into power-generation-ready fuel. For operations in a remote or harsh environment, SoEnergy leverages extensive logistics and engineering support to simplify, and solve, the challenge. SoEnergy has deployed this technology in projects in Colombia and Ecuador.

(continued)

TABLE A.1, *continued*

DNV GL	DNV GL has proposed alternatives to gas flaring. DNV GL has developed a methodology that uses gas flowrate and distance to market to select the most appropriate technical solutions on a case-by-case basis. The methodology could present new revenue opportunities, particularly for smaller-scale applications for operators, while helping them to reduce emissions and stay ahead of regulatory requirements.
	Though some solutions might be immature for near-term implementation, current applications—such as micro LNG, compressed natural gas, NGHs, and conversion methods—can deliver significant benefits, and are proving to do so in some cases in certain markets like North America.
Crusoe	Crusoe Energy Systems provides oil and gas companies with a fast, low-cost, and simple solution to natural gas flaring. As the energy industry struggles with pipeline constraints and increasing regulations around flaring and combusting, Crusoe's service lets operators preserve or regain regulatory compliance, maintain existing production, and facilitate future development. Crusoe's Digital Flare Mitigation™ (DFM) systems convert otherwise wasted natural gas into electricity to power energy-intensive computing right at the wellsite.
Mitsubishi Corporation	Basra Gas Company, 2013 (Iraq): Since operations began, the company has been recovering and refining flare gas from the three oil fields in southern Iraq (Rumaila, West Qurna 1, Zubair), and selling gas for power generation, LPG, and condensates both in Iraq and abroad.
	Sakhalin 2 Project (Russia): An integrated oil and gas development business that produces crude oil from the oil field in the north of Sakhalin and also liquefies natural gas produced from the gas fields in the north of Sakhalin.
	Montney shale gas development project (Canada): "Building a natural gas value chain in Canada," the project stretches from upstream resource development to LNG production, export, and sales. In terms of upstream businesses, the company works with Ovintiv Inc. on developing and producing shale gas from Montney Formation in the area near Dawson Creek, British Columbia.

Source: Based on company websites.
Note: Btu = British thermal unit; capex = capital expenditures; LNG = liquefied natural gas; LPG = liquefied petroleum gas; NGH = natural gas hydrate; NGL = natural gas liquids; opex = operating expenditures.

Detailed Technical Assumptions on Power Generation Efficiency, Capital Expenditures, and Operating Expenditures

TABLE B.1 **General inputs**

YEARS		0	1	2	3	4	5	6	7
Flare size of the project									
Small	mmscf/d	1							
Medium	mmscf/d	5							
Large	mmscf/d	10							
Routine flare		85%							
Routine flare size									
Small	mmscf/d	0.850							
Medium	mmscf/d	4.250							
Large	mmscf/d	8.500							
Decline rate (%)		0	5	5	5	5	5	5	5
Cumulative (%)		0.00	5.00	9.75	14.26	18.55	22.62	26.49	30.17
Routine flare profile									
Small	mmscf/d	0.850	0.808	0.767	0.729	0.692	0.658	0.625	0.594
Medium	mmscf/d	4.250	4.038	3.836	3.644	3.462	3.289	3.124	2.968
Large	mmscf/d	8.500	8.075	7.671	7.288	6.923	6.577	6.248	5.936
AG price paid to operator	US$/mscf	0.25	(in scenarios where FMR developer does pay for AG)						
Discount rate (real)		10%							
Depreciation period	years	5							
Tax rate (corporate)		0%							
Annual inflation		0%							

Source: World Bank.
Note: AG = associated gas; FMR = flaring and methane reduction; mmscf/d = million standard cubic feet per day; mscf = thousand standard cubic feet.

POWER GENERATION

TABLE B.2 **Inputs for efficiency**

SIZE	FLARE SIZE (mmscf/d)	ROUTINE FLARE SIZE (mmscf/d)[a]	FUEL GAS RATIO[a]	EFFICIENCY (Btu/kWh)	LHV FUEL GAS (Btu/scf)	GAS PER MW (mmscf/d)	POWER PLANT SIZE (MW)[b]	PARASITIC LOAD	OBSERVATION
Small	1	0.85	98%	11,000	1,050	0.2514	3.31	3.0%	Recip engines with capacity to run on "raw" gas.[c] Simple cycle gas turbines have a lower efficiency (11,500 Btu/kWh–12,500 Btu/kWh).
Medium	5	4.25	93%	8,900	996	0.2144	18.44	2.5%	Recip engines with capacity to run on "semi-raw" gas.[d] Simple cycle gas turbines have a lower efficiency (10,000 Btu/kWh–11,000 Btu/kWh).
Large	10	8.50	93%	7,900	996	0.1903	41.54	2.0%	Recip engines with capacity to run on "semi-raw" gas.[d] Simple cycle gas turbines have a lower efficiency (9,000 Btu/kWh–10,000 Btu/kWh).

Source: World Bank.
Note: All efficiencies are for simple cycle power plants. Btu = British thermal unit; kWh = kilowatt-hour; LHV = lower heating value; mmscf/d = million standard cubic feet per day; MW = megawatt; scf = standard cubic feet.
a. "Small" size assumes no "dew point plant" (only shrinkage due to compression). "Medium" and "large" assume a minimum dew point process (only to extract heavier hydrocarbons to avoid "knocking").
b. Assuming 85 percent of the flare gas is routine flaring.
c. There are more efficient recip engines on the market, but these do not have the ability to run on "raw" (untreated) gas.
d. Typically these recip engines will run on raw gas, but typically there will be some dew point adjustment (to eliminate heavier hydrocarbons to avoid "knocking").

TABLE B.3 **Inputs for capital expenditures**

SIZE	FLARE SIZE (mmscf/d)	ROUTINE FLARE SIZE (mmscf/d)[a]	POWER PLANT SIZE (MW)	CAPEX (US$/MW)	OBSERVATION
Small	1	0.85	3.31	$2,500,000	Gas Modular Power Units (G-MPUs) in Ecuador (2 MW–7 MW each G-MPU) were installed at a cost of US$2,500 per kW (CPF BL15, LPF, Paka Sur, Yamanunka, Cuyabeno, Sacha Norte 2, Sacha Central, Sacha Sur, etc.). This includes tie-ins, compression, and electrical interconnection. Assumes plug-and-plug equipment with minimal civil works. Recent projects point to possibly US$2.2m/MWh with modular equipment on wheels.
Medium	5	4.25	18.44	$1,800,000	
Large	10	8.50	41.54	$1,500,000	El Morro Power Project in Colombia (60 MW) was installed at a cost of US$1,000 per kW (including a 17.5 km transmission line). This nevertheless did not include compression, which is why we recommend using US$1,200/kW.

Source: World Bank.
Note: capex = capital expenditures; kWh = kilowatt-hour; mmscf/d = million standard cubic feet per day; MW = megawatt.
a. "Small" size assumes no "dew point plant" (only shrinkage due to compression). "Medium" and "large" assume a minimum dew point process (only to extract heavier hydrocarbons to avoid "knocking").

Inputs for operating expenditures

WORK/SCOPE	SMALL US$/kWh	MEDIUM US$/kWh	LARGE US$/kWh	OBSERVATION
Scheduled maintenance	$0.0040	$0.0036	$0.0034	
Provision for major overhaul	30%	30%	30%	% of scheduled maintenance
Provision for major overhaul	$0.0012	$0.0011	$0.0010	
Unscheduled maintenance	25%	25%	25%	% of scheduled maintenance
Unscheduled maintenance	$0.0010	$0.0009	$0.0009	
Lubricants/consumables	$0.0020	$0.0018	$0.0017	
Balance of plant maintenance	$0.0020	$0.0018	$0.0017	
Variable O&M	$0.0102	$0.0092	$0.0087	
Fixed O&M cost	$0.0250	$0.0180	$0.0100	24/7, day and night shift and 14/14 day shift
Total O&M	$0.0352	$0.0272	$0.0187	
Capex (US$/kW)	$2,500	$1,800	$1,500	
Utilization factor (availability x load factor)	90%	90%	90%	Assuming a "must run" plant
Opex (US$/kW)	$277.52	$214.29	$147.60	
Opex (%/capex)	11.10%	11.90%	9.84%	

Source: World Bank.
Note: Numbers in table B.4 are averages derived from a large sample of O&M negotiated signed contracts. capex = capital expenditures; kWh = kilowatt-hour; opex = operating expenditures; O&M = operations and maintenance.

TABLE B.5 **Power (external use) base case assumptions**

MODEL VARIABLE	FLARE SIZE	UNIT OF MEASURE	VALUE	COMMENTS
Fuel gas ratio (after elimination of liquids)	Small		98%	Compression only for small projects (to keep capex low); no dewpointing
	Medium		93%	Includes dewpointing
	Large		93%	Includes dewpointing
Parasitic gas usage (needed to run power plant)	Small		3.0%	Small plants tend to have higher parasitic load
	Medium		2.5%	
	Large		2.0%	
Conversion heat content to power	Small	Btu/kWh	11,000	Based on real project data
	Medium	Btu/kWh	8,900	Based on real project data
	Large	Btu/kWh	7,900	Based on real project data
Gas energy content	Small	Btu/scf	1,050	
	Medium	Btu/scf	996	
	Large	Btu/scf	996	
Fuel gas needed for 1 MW powergen	Small	mmscf/d	0.251	
	Medium		0.203	
	Large		0.181	
Capex per 1 MW powergen	Small	US$/MW	2,500,000	Includes provisions for major overhaul and unscheduled maintenance
	Medium	US$/MW	1,800,000	Includes provisions for major overhaul, unscheduled maintenance, and dew point
	Large	US$/MW	1,500,000	Includes provisions for major overhaul, unscheduled maintenance, and dew point
Capex for transmission line (overhead)	Small	US$/km	60,000	
	Medium	US$/km	100,000	Requires steel towers, substations
	Large	US$/km	100,000	Requires steel towers, substations
Distance to grid	Small	km	5	
	Medium	km	10	
	Large	km	30	
Opex for generation	Small	% of capex	11.1%	Includes provisions for major overhaul and unscheduled maintenance
	Medium	% of capex	11.9%	
	Large	% of capex	9.8%	
Opex for transmission	% of capex	2.0%		
Electricity price (wholesale)	Small	US$/kWh	$0.08	
	Medium	US$/kWh	$0.08	
	Large	US$/kWh	$0.08	
	Small	US$/kWh	$0.08	
	Medium	US$/kWh	$0.08	
	Large	US$/kWh	$0.08	
AG price paid to operator	US$/mscf	$0.25		
Power plant availability	% days in year	94%		

Source: World Bank.
Note: Base case project duration from capex completion is modeled at 7 years. AG = associated gas; Btu = British thermal unit; capex = capital expenditures; km = kilometer; kWh = kilowatt-hour; mmscf/d = million standard cubic feet per day; mscf = thousand standard cubic feet; opex = operating expenditures; scf = standard cubic feet.

TABLE B.6 Power (external use) base case calculations

PARAMETER	FLARE SIZE	UNIT	YEAR							
			0	1	2	3	4	5	6	7
Gas flared (routine)	Small	mmscf/d	0.850	0.808	0.767	0.729	0.692	0.658	0.625	0.594
	Medium	mmscf/d	4.250	4.038	3.836	3.644	3.462	3.289	3.124	2.968
	Large	mmscf/d	8.500	8.075	7.671	7.288	6.923	6.577	6.248	5.936
Fuel gas available (per day)	Small	mmscf/d	0.833	0.791	0.752	0.714	0.678	0.645	0.612	0.582
	Medium	mmscf/d	3.953	3.755	3.567	3.389	3.219	3.058	2.905	2.760
	Large	mmscf/d	7.905	7.510	7.134	6.778	6.439	6.117	5.811	5.520
Generation capacity installed (sized on 3-year fuel gas available)	Small	MW	2.9							
	Medium	MW	16.7							
	Large	MW	37.6							
Capex—generation	Small	US$	7,250,000							
	Medium	US$	30,060,000							
	Large	US$	56,400,000							
Capex—transmission	Small	US$	300,000							
	Medium	US$	1,000,000							
	Large	US$	3,000,000							
Power produced	Small	kWh		22,688,453	22,688,453	22,688,453	21,554,030	20,476,329	19,452,512	18,479,887
	Medium	kWh		133,741,848	133,741,848	133,741,848	127,054,756	120,702,018	114,666,917	108,933,571
	Large	kWh		302,887,738	302,887,738	302,887,738	287,743,351	273,356,183	259,688,374	246,703,956
Revenue	Small	US$		1,815,076	1,815,076	1,815,076	1,724,322	1,638,106	1,556,201	1,478,391
	Medium	US$		10,699,348	10,699,348	10,699,348	10,164,380	9,656,161	9,173,353	8,714,686
	Large	US$		24,231,019	24,231,019	24,231,019	23,019,468	21,868,495	20,775,070	19,736,316
Opex—generation	Small	US$		804,799	804,799	804,799	804,799	804,799	804,799	804,799
	Medium	US$		3,578,595	3,578,595	3,578,595	3,578,595	3,578,595	3,578,595	3,578,595
	Large	US$		5,549,623	5,549,623	5,549,623	5,549,623	5,549,623	5,549,623	5,549,623
Opex—transmission	Small	US$		6,000	6,000	6,000	6,000	6,000	6,000	6,000
	Medium	US$		20,000	20,000	20,000	20,000	20,000	20,000	20,000
	Large	US$		60,000	60,000	60,000	60,000	60,000	60,000	60,000

(continued)

TABLE B.6, *continued*

PARAMETER	FLARE SIZE	UNIT	YEAR							
			0	1	2	3	4	5	6	7
Opex—purchase of AG	Small	US$		61,260	61,260	61,260	58,197	55,287	52,523	49,897
	Medium	US$		290,672	290,672	290,672	276,139	262,332	249,215	236,754
	Large	US$		581,344	581,344	581,344	552,277	524,663	498,430	473,509
Capex to depreciate	Small	US$	7,550,000							
	Medium	US$	31,060,000							
	Large	US$	59,400,000							
Depreciation	Small	US$		1,510,000	1,510,000	1,510,000	1,510,000	1,510,000	0	0
	Medium	US$		6,212,000	6,212,000	6,212,000	6,212,000	6,212,000	0	0
	Large	US$		11,880,000	11,880,000	11,880,000	11,880,000	11,880,000	0	0
Corporate tax base	Small	US$		(566,982)	(566,982)	(566,982)	(654,673)	(737,980)	692,880	617,696
	Medium	US$		598,081	598,081	598,081	77,647	(416,765)	5,325,543	4,879,337
	Large	US$		6,160,051	6,160,051	6,160,051	4,977,568	3,854,208	14,667,017	13,653,185
Corporate tax with loss carry-forward and no other taxable income	Small taxable income			(566,982)	(1,133,965)	(1,700,947)	(2,355,621)	(3,093,600)	(2,400,720)	(1,783,025)
	Small tax paid			0	0	0	0	0	0	0
	Medium taxable income			598,081	1,196,162	1,794,242	1,871,889	1,455,124	6,780,668	11,660,004
	Medium tax paid			0	0	0	0	0	0	0
	Large taxable income			6,160,051	12,320,103	18,480,154	23,457,722	27,311,930	41,978,947	55,632,131
	Large tax paid			0	0	0	0	0	0	0

(continued)

TABLE B.6, *continued*

PARAMETER	FLARE SIZE	UNIT	YEAR							
			0	1	2	3	4	5	6	7
Net cashflows	Small	US$	(7,550,000)	943,018	943,018	943,018	855,327	772,020	692,880	617,696
	Medium	US$	(31,060,000)	6,810,081	6,810,081	6,810,081	6,289,647	5,795,235	5,325,543	4,879,337
	Large	US$	(59,400,000)	18,040,051	18,040,051	18,040,051	16,857,568	15,734,208	14,667,017	13,653,185
NPV			7 years		IRR	7 years	5 years	10 years		
	Small	US$m	(3.1)			−7%	−16%	−1%		
	Medium	US$m	(0.7)			9%	2%	14%		
	Large	US$m	20.0			21%	14%	25%		

Source: World Bank.

Note: Base case project duration from capex completion is modeled at 7 years. IRR sensitivity is modeled at 5 and 10 years from capex completion. AG = associated gas; capex = capital expenditures; IRR = internal rate of return; kWh = kilowatt-hour; mmscf/d = million standard cubic feet per day; NPV = net present value; opex = operating expenditures.

TABLE B.7 **Power (on-site) base case assumptions**

MODEL VARIABLE	FLARE SIZE	UNIT OF MEASURE	VALUE	COMMENTS
Fuel gas ratio (after elimination of liquids)	Small		98%	Compression only for small projects (to keep capex low); no dewpointing
	Medium		93%	Includes dewpointing
	Large		93%	Includes dewpointing
Parasitic gas usage (needed to run power plant)	Small		3.0%	Small plants tend to have higher parasitic load
	Medium		2.5%	
	Large		2.0%	
Conversion heat content to power	Small	Btu/kWh	11,000	Based on real project data
	Medium	Btu/kWh	8,900	Based on real project data
	Large	Btu/kWh	7,900	Based on real project data
Gas energy content	Small	Btu/scf	1,050	
	Medium	Btu/scf	996	
	Large	Btu/scf	996	
Fuel gas needed for 1 MW powergen	Small	mmscf/d	0.251	
	Medium		0.203	
	Large		0.181	
Capex per 1 MW powergen	Small	US$/MW	2,500,000	Includes gas compression equipment, power generation equipment and back-end connection
	Medium	US$/MW	1,800,000	Also includes dew point
	Large	US$/MW	1,500,000	Also includes dew point
Capex for transmission line (overhead)	Small	US$/km	0	
	Medium	US$/km	0	Requires steel towers, substations
	Large	US$/km	0	Requires steel towers, substations
Distance to grid	Small	km	0	
	Medium	km	0	
	Large	km	0	
Opex for generation	Small	% of capex	11.1%	Includes provisions for major overhaul and unscheduled maintenance
	Medium	% of capex	11.9%	Includes provisions for major overhaul and unscheduled maintenance
	Large	% of capex	9.8%	Includes provisions for major overhaul and unscheduled maintenance
Opex for transmission		US$	0.0%	
Electricity price (wholesale)	Small	US$/kWh	$0.100	Higher than in external case as project replaces expensive diesel generation
	Medium	US$/kWh	$0.085	Decreases with increasing project size as oil company is able to extract better terms
	Large	US$/kWh	$0.070	Decreases with increasing project size as oil company is able to extract better terms
	Small	US$/kWh	$0.100	
	Medium	US$/kWh	$0.085	
	Large	US$/kWh	$0.070	
AG price paid to operator		US$/mscf	0.00	Operator gives AG for free but obtains cheaper power than the diesel gen alternative
Power plant availability		% days in year	94%	

Source: World Bank.
Note: Base case project duration from capex completion is modeled at 7 years. AG = associated gas; Btu = British thermal unit; capex = capital expenditures; km = kilometer; kWh = kilowatt-hour; mmscf/d = million standard cubic feet per day; mscf = thousand standard cubic feet; MW = megawatt; opex = operating expenditures.

TABLE B.8 **Power (on-site) base case calculations**

PARAMETER	FLARE SIZE	UNIT	0	1	2	3	4	5	6	7
Gas flared (routine)	Small	mmscf/d	0.850	0.808	0.767	0.729	0.692	0.658	0.625	0.594
	Medium	mmscf/d	4.250	4.038	3.836	3.644	3.462	3.289	3.124	2.968
	Large	mmscf/d	8.500	8.075	7.671	7.288	6.923	6.577	6.248	5.936
Fuel gas available (per day)	Small	mmscf/d	0.833	0.791	0.752	0.714	0.678	0.645	0.612	0.582
	Medium	mmscf/d	3.953	3.755	3.567	3.389	3.219	3.058	2.905	2.760
	Large	mmscf/d	7.905	7.510	7.134	6.778	6.439	6.117	5.811	5.520
Generation capacity installed	Small	MW	2.9							
	Medium	MW	16.7			Sized on year 3 fuel gas available				
	Large	MW	37.6							
Capex—generation	Small	US$	7,250,000							
	Medium	US$	30,060,000							
	Large	US$	56,400,000							
Capex—transmission	Small	US$	300,000							
	Medium	US$	1,000,000							
	Large	US$	3,000,000							
Power produced	Small	kWh		22,688,453	22,688,453	22,688,453	21,554,030	20,476,329	19,452,512	18,479,887
	Medium	kWh		133,741,848	133,741,848	133,741,848	127,054,756	120,702,018	114,666,917	108,933,571
	Large	kWh		302,887,738	302,887,738	302,887,738	287,743,351	273,356,183	259,688,374	246,703,956
Revenues	Small	US$		1,815,076	1,815,076	1,815,076	1,724,322	1,638,106	1,556,201	1,478,391
	Medium	US$		10,699,348	10,699,348	10,699,348	10,164,380	9,656,161	9,173,353	8,714,686
	Large	US$		24,231,019	24,231,019	24,231,019	23,019,468	21,868,495	20,775,070	19,736,316
Opex—generation	Small	US$		804,799	804,799	804,799	804,799	804,799	804,799	804,799
	Medium	US$		3,578,595	3,578,595	3,578,595	3,578,595	3,578,595	3,578,595	3,578,595
	Large	US$		5,549,623	5,549,623	5,549,623	5,549,623	5,549,623	5,549,623	5,549,623
Opex—transmission	Small	US$		6,000	6,000	6,000	6,000	6,000	6,000	6,000
	Medium	US$		20,000	20,000	20,000	20,000	20,000	20,000	20,000
	Large	US$		60,000	60,000	60,000	60,000	60,000	60,000	60,000
Opex—purchase of AG	Small	US$		61,260	61,260	61,260	58,197	55,287	52,523	49,897
	Medium	US$		290,672	290,672	290,672	276,139	262,332	249,215	236,754
	Large	US$		581,344	581,344	581,344	552,277	524,663	498,430	473,509

(continued)

TABLE B.8, *continued*

PARAMETER	FLARE SIZE	UNIT	0	1	2	3	4	5	6	7
Capex to depreciate	Small	US$	7,550,000							
	Medium	US$	31,060,000							
	Large	US$	59,400,000							
Depreciation	Small	US$		1,510,000	1,510,000	1,510,000	1,510,000	1,510,000	0	0
	Medium	US$		6,212,000	6,212,000	6,212,000	6,212,000	6,212,000	0	0
	Large	US$		11,880,000	11,880,000	11,880,000	11,880,000	11,880,000	0	0
Corporate tax base	Small	US$		(566,982)	(566,982)	(566,982)	(654,673)	(737,980)	692,880	617,696
	Medium	US$		598,081	598,081	598,081	77,647	(416,765)	5,325,543	4,879,337
	Large	US$		6,160,051	6,160,051	6,160,051	4,977,568	3,854,208	14,667,017	13,653,185
Corporate tax with loss carry-forward and no other taxable income	Small taxable income			(566,982)	(1,133,965)	(1,700,947)	(2,355,621)	(3,093,600)	(2,400,720)	(1,783,025)
	Small tax paid			0	0	0	0	0	0	0
	Medium taxable income			598,081	1,196,162	1,794,242	1,871,889	1,455,124	6,780,668	11,660,004
	Medium tax paid			0	0	0	0	0	0	0
	Large taxable income			6,160,051	12,320,103	18,480,154	23,457,722	27,311,930	41,978,947	55,632,131
	Large tax paid			0	0	0	0	0	0	0
Net cashflows	Small	US$	(7,550,000)	943,018	943,018	943,018	855,327	772,020	692,880	617,696
	Medium	US$	(31,060,000)	6,810,081	6,810,081	6,810,081	6,289,647	5,795,235	5,325,543	4,879,337
	Large	US$	(59,400,000)	18,040,051	18,040,051	18,040,051	16,857,568	15,734,208	14,667,017	13,653,185

			NPV (7 years)	IRR		
				5 years	7 years	10 years
NPV	Small	US$m	(3.1)	−16%	−7%	−1%
	Medium	US$m	(0.7)	2%	9%	14%
	Large	US$m	20.0	14%	21%	25%

Source: World Bank.

Note: Base case project duration from capex completion is modeled at 7 years. IRR sensitivity is modeled at 5 and 10 years from capex completion. AG = associated gas; Btu = British thermal unit; capex = capital expenditures; IRR = internal rate of return; km = kilometer; kWh = kilowatt-hour; mmscf/d = million standard cubic feet per day; MW = megawatt; NPV = net present value; opex = operating expenditures; scf = standard cubic feet.

TABLE B.9 **Pipe to pipe base case assumptions**

MODEL VARIABLE	FLARE SIZE	UNIT OF MEASURE	VALUE	COMMENTS
Fuel gas ratio			90%	10% less because of LPG extraction (and no sale of LPG)
Parasitic gas usage			13%	Gas burnt to extract LPG (7%), run dew point (3%), and compress (3%)
Capex for liquids removal, dew point, and compression	Small	US$/mmscf/d	3,280,000	
	Medium	US$/mmscf/d	1,780,000	
	Large	US$/mmscf/d	1,580,000	
Capex for pipeline	Small	US$/km	300,000	
	Medium	US$/km	450,000	
	Large	US$/km	600,000	
Distance to existing pipeline network	Small	km	1	
	Medium	km	3	
	Large	km	5	
Opex		% total capex	3.5%	
Tolling fee	Small	US$/mscf	2.00	For both dew point / compression and pipeline transport
	Medium	US$/mscf	2.00	
	Large	US$/mscf	2.00	
Compressor / dew point availability		% days in year	98.0%	
LPG price		US$/mscf	0.00	

Source: World Bank.

Note: Base case project duration from capex completion is modeled at 7 years. capex = capital expenditures; km = kilometer; LPG = liquefied petroleum gas; mmscf/d = million standard cubic feet per day; mscf = thousand standard cubic feet; opex = operating expenditures.

TABLE B.10 Pipe to pipe base case calculations

PARAMETER	FLARE SIZE	UNIT	YEAR							
			0	1	2	3	4	5	6	7
Fuel gas treated and compressed (per day)	Small	mmscf/d	0.765	0.727	0.690	0.656	0.623	0.592	0.562	0.534
	Medium	mmscf/d	3.825	3.634	3.452	3.279	3.115	2.960	2.812	2.671
	Large	mmscf/d	7.650	7.268	6.904	6.559	6.231	5.919	5.623	5.342
Capex—liquids removal, dew point and compression	Small	US$	2,296,000							
	Medium	US$	5,874,000							
	Large	US$	10,428,000							
Capex—pipeline	Small	US$	300,000							
	Medium	US$	1,350,000							
	Large	US$	3,000,000							
Revenues from pipeline	Small	US$		408,226	408,226	408,226	387,815	368,424	350,003	332,502
	Medium	US$		2,041,129	2,041,129	2,041,129	1,939,073	1,842,119	1,750,013	1,662,512
	Large	US$		4,082,258	4,082,258	4,082,258	3,878,145	3,684,238	3,500,026	3,325,025
Revenues from sale of LPG	Small	US$		0	0	0	0	0	0	0
	Medium	US$		0	0	0	0	0	0	0
	Large	US$		0	0	0	0	0	0	0
Opex	Small	US$		90,860	90,860	90,860	90,860	90,860	90,860	90,860
	Medium	US$		252,840	252,840	252,840	252,840	252,840	252,840	252,840
	Large	US$		469,980	469,980	469,980	469,980	469,980	469,980	469,980
Capex to depreciate	Small	US$	2,596,000							
	Medium	US$	7,224,000							
	Large	US$	13,428,000							
Depreciation	Small	US$		519,200	519,200	519,200	519,200	519,200	0	0
	Medium	US$		1,444,800	1,444,800	1,444,800	1,444,800	1,444,800	0	0
	Large	US$		2,685,600	2,685,600	2,685,600	2,685,600	2,685,600	0	0

(continued)

TABLE B.10, *continued*

PARAMETER	FLARE SIZE	UNIT	0	1	2	3	4	5	6	7
Corporate tax base	Small	US$		(201,834)	(201,834)	(201,834)	(222,245)	(241,636)	259,143	241,642
	Medium	US$		343,489	343,489	343,489	241,433	144,479	1,497,173	1,409,672
	Large	US$		926,678	926,678	926,678	722,565	528,658	3,030,046	2,855,045
Corporate tax with loss carry-forward and no other taxable income	Small taxable income			(201,834)	(403,668)	(605,503)	(827,748)	(1,069,384)	(810,242)	(568,599)
	Small tax paid			0	0	0	0	0	0	0
	Medium taxable income			343,489	686,978	1,030,467	1,271,899	1,416,378	2,913,551	4,323,223
	Medium tax paid			0	0	0	0	0	0	0
	Large taxable income			926,678	1,853,356	2,780,034	3,502,599	4,031,257	7,061,302	9,916,347
	Large tax paid			0	0	0	0	0	0	0
Net cashflow	Small	US$	(2,596,000)	317,366	317,366	317,366	296,955	277,564	259,143	241,642
	Medium	US$	(7,224,000)	1,788,289	1,788,289	1,788,289	1,686,233	1,589,279	1,497,173	1,409,672
	Large	US$	(13,428,000)	3,612,278	3,612,278	3,612,278	3,408,165	3,214,258	3,030,046	2,855,045

PARAMETER	FLARE SIZE	UNIT	7 years		IRR		7 years	5 years	10 years
NPV	Small	US$m	(1.1)				−6%	−16%	0%
	Medium	US$m	0.8				14%	6%	18%
	Large	US$m	2.8				17%	10%	21%

Source: World Bank.

Note: Base case project duration from capex completion is modeled at 7 years. IRR sensitivity is modeled at 5 and 10 years from capex completion. capex = capital expenditures; IRR = internal rate of return; LPG = liquefied petroleum gas; mmscf/d = million standard cubic feet per day; NPV = net present value; opex = operating expenditures.

TABLE B.11 **Pipe to gas processing plant base case assumptions**

MODEL VARIABLE	FLARE SIZE	UNIT OF MEASURE	VALUE	COMMENTS
Parasitic gas usage			3%	Compression only
Capex for compression	Small	US$/mmscf/d	600,000	
	Medium	US$/mmscf/d	350,000	
	Large	US$/mmscf/d	310,000	
Capex for pipeline	Small	US$/km	300,000	
	Medium	US$/km	450,000	
	Large	US$/km	600,000	
Distance to GPP	Small	km	5	
	Medium	km	15	
	Large	km	20	
Opex		% total capex	3.5%	
Tolling fee	Small	US$/mscf	1.50	For compression and pipeline transport
	Medium	US$/mscf	1.50	
	Large	US$/mscf	1.50	
Compressor / dew point availability		% days in year	98.0%	

Source: World Bank.

Note: Base case project duration from capex completion is modeled at 7 years. capex = capital expenditures; GPP = gas processing plant; km = kilometer; mmscf/d = million standard cubic feet per day; mscf = thousand standard cubic feet; opex = operating expenditures.

TABLE B.12 **Pipe to gas processing plant base case calculations**

PARAMETER	FLARE SIZE	UNIT	0	1	2	3	4	5	6	7
						YEAR				
Fuel gas treated and compressed (per day)	Small	mmscf/d	0.850	0.808	0.767	0.729	0.692	0.658	0.625	0.594
	Medium	mmscf/d	4.250	4.038	3.836	3.644	3.462	3.289	3.124	2.968
	Large	mmscf/d	8.500	8.075	7.671	7.288	6.923	6.577	6.248	5.936
Capex—dew point and compression	Small	US$	480,000							
	Medium	US$	1,295,000							
	Large	US$	2,263,000							
Capex—pipeline	Small	US$	1,500,000							
	Medium	US$	6,750,000							
	Large	US$	12,000,000							
Revenues	Small	US$		379,290	379,290	379,290	360,326	342,309	325,194	308,934
	Medium	US$		1,896,451	1,896,451	1,896,451	1,801,629	1,711,547	1,625,970	1,544,671
	Large	US$		3,792,902	3,792,902	3,792,902	3,603,257	3,423,094	3,251,940	3,089,343
Opex	Small	US$		69,300	69,300	69,300	69,300	69,300	69,300	69,300
	Medium	US$		281,575	281,575	281,575	281,575	281,575	281,575	281,575
	Large	US$		499,205	499,205	499,205	499,205	499,205	499,205	499,205
Capex to depreciate	Small	US$	1,980,000							
	Medium	US$	8,045,000							
	Large	US$	14,263,000							
Depreciation	Small	US$		396,000	396,000	396,000	396,000	396,000	0	0
	Medium	US$		1,609,000	1,609,000	1,609,000	1,609,000	1,609,000	0	0
	Large	US$		2,852,600	2,852,600	2,852,600	2,852,600	2,852,600	0	0
Corporate tax base	Small	US$		(86,010)	(86,010)	(86,010)	(104,974)	(122,991)	255,894	239,634
	Medium	US$		5,876	5,876	5,876	(88,946)	(179,028)	1,344,395	1,263,096
	Large	US$		441,097	441,097	441,097	251,452	71,289	2,752,735	2,590,138

(continued)

TABLE B.12, *continued*

PARAMETER	FLARE SIZE	UNIT	YEAR 0	1	2	3	4	5	6	7
Corporate tax with loss carry-forward and no other taxable income	Small taxable income			(86,010)	(172,020)	(258,029)	(363,004)	(485,994)	(230,100)	9,534
	Small tax paid			0	0	0	0	0	0	0
	Medium taxable income			5,876	11,752	17,629	(71,318)	(250,345)	1,094,049	2,357,146
	Medium tax paid			0	0	0	0	0	0	0
	Large taxable income			441,097	882,195	1,323,292	1,574,745	1,646,034	4,398,769	6,988,907
	Large tax paid			0	0	0	0	0	0	0
Net cashflows	Small	US$	(1,980,000)	309,990	309,990	309,990	291,026	273,009	255,894	239,634
	Medium	US$	(8,045,000)	1,614,876	1,614,876	1,614,876	1,520,054	1,429,972	1,344,395	1,263,096
	Large	US$	(14,263,000)	3,293,697	3,293,697	3,293,697	3,104,052	2,923,889	2,752,735	2,590,138

NPV			7 years		IRR		7 years	5 years	10 years
Small	US$m	(0.5)					0%	−9%	6%
Medium	US$m	(0.6)					7%	−1%	12%
Large	US$m	0.7					12%	4%	16%

Source: World Bank.

Note: Base case project duration from capex completion is modeled at 7 years. IRR sensitivity is modeled at 5 and 10 years from capex completion. capex = capital expenditures; IRR = internal rate of return; mmscf/d = million standard cubic feet per day; NPV = net present value; opex = operating expenditures.

TABLE B.13 Compressed natural gas to gas processing plant base case assumptions

MODEL VARIABLE	FLARE SIZE	UNIT OF MEASURE	VALUE	COMMENTS
Capex (LPG extraction, dew point, compression, trucks)	Small	US$/mmscf/d	8,750,000	
	Medium	US$/mmscf/d	4,250,000	
	Large	US$/mmscf/d	3,700,000	
Opex		% total capex	13%	Including for trucks, assuming they do not travel beyond a 150 km radius
Fuel gas ratio			90%	After compression and elimination of liquids
Parasitic gas usage			14%	7% for liquid removal, 3% for dewpointing, and 4% for compression to 3,000 psi (higher pressure requires more energy)
CNG price		US$/mscf	6.00	
AG purchase price		US$/mscf	0.25	
CNG availability		% days in year	94.0%	Compression and transportation downtime
LPG price		US$/mscf	0.00	

Source: World Bank.

Note: Base case project duration from capex completion is modeled at 7 years. AG = associated gas; capex = capital expenditures; CNG = compressed natural gas; km = kilometer; LPG = liquefied petroleum gas; mmscf/d = million standard cubic feet per day; mscf = thousand standard cubic feet; opex = operating expenditures; psi = pounds per square inch.

TABLE B.14 Compressed natural gas to gas processing plant base case calculations

PARAMETER	FLARE SIZE	UNIT	YEAR							
			0	1	2	3	4	5	6	7
Gas flared (routine)	Small	mmscf/d	0.850	0.808	0.767	0.729	0.692	0.658	0.625	0.594
	Medium	mmscf/d	4.250	4.038	3.836	3.644	3.462	3.289	3.124	2.968
	Large	mmscf/d	8.500	8.075	7.671	7.288	6.923	6.577	6.248	5.936
Fuel gas treated and compressed (per day)	Small	mmscf/d	0.765	0.727	0.690	0.656	0.623	0.592	0.562	0.534
	Medium	mmscf/d	3.825	3.634	3.452	3.279	3.115	2.960	2.812	2.671
	Large	mmscf/d	7.650	7.268	6.904	6.559	6.231	5.919	5.623	5.342
Capex	Small	US$	6,125,000							
	Medium	US$	14,025,000							
	Large	US$	24,420,000							
Revenues from sale of CNG	Small	US$		1,161,188	1,161,188	1,161,188	1,103,129	1,047,972	995,574	945,795
	Medium	US$		5,805,942	5,805,942	5,805,942	5,515,645	5,239,862	4,977,869	4,728,976
	Large	US$		11,611,884	11,611,884	11,611,884	11,031,289	10,479,725	9,955,739	9,457,952
Revenues from sale of LPG	Small	US$		0	0	0	0	0	0	0
	Medium	US$		0	0	0	0	0	0	0
	Large	US$		0	0	0	0	0	0	0
Opex	Small	US$		796,250	796,250	796,250	796,250	796,250	796,250	796,250
	Medium	US$		1,823,250	1,823,250	1,823,250	1,823,250	1,823,250	1,823,250	1,823,250
	Large	US$		3,174,600	3,174,600	3,174,600	3,174,600	3,174,600	3,174,600	3,174,600
Purchase of AG	Small	US$		56,259	56,259	56,259	53,446	50,774	48,235	45,823
	Medium	US$		281,296	281,296	281,296	267,231	253,869	241,176	229,117
	Large	US$		562,591	562,591	562,591	534,462	507,739	482,352	458,234
Capex to depreciate	Small	US$	6,125,000							
	Medium	US$	14,025,000							
	Large	US$	24,420,000							
Depreciation	Small	US$		1,225,000	1,225,000	1,225,000	1,225,000	1,225,000	0	0
	Medium	US$		2,805,000	2,805,000	2,805,000	2,805,000	2,805,000	0	0
	Large	US$		4,884,000	4,884,000	4,884,000	4,884,000	4,884,000	0	0

(continued)

TABLE B.14, *continued*

PARAMETER	FLARE SIZE	UNIT	YEAR 0	1	2	3	4	5	6	7
Corporate tax base	Small	US$	(916,321)	(916,321)	(916,321)	(916,321)	(971,567)	(1,024,051)	151,089	103,722
	Medium	US$	896,396	896,396	896,396	896,396	620,164	357,743	2,913,443	2,676,609
	Large	US$	2,990,692	2,990,692	2,990,692	2,990,692	2,438,228	1,913,386	6,298,787	5,825,118
Corporate tax with loss carry-forward and no other taxable income	Small taxable income			(916,321)	(1,832,642)	(2,748,962)	(3,720,530)	(4,744,581)	(4,593,492)	(4,489,770)
	Small tax paid			0	0	0	0	0	3	0
	Medium taxable income			896,396	1,792,792	2,689,188	3,309,352	3,667,095	6,580,539	9,257,148
	Medium tax paid			0	0	0	0	0	0	0
	Large taxable income			2,990,692	5,981,385	8,972,077	11,410,304	13,323,691	19,622,478	25,447,595
	Large tax paid			0	0	0	0	0	0	0
Net cashflow	Small	US$	(6,125,000)	308,679	308,679	308,679	253,433	200,949	151,086	103,722
	Medium	US$	(14,025,000)	3,701,396	3,701,396	3,701,396	3,425,164	3,162,743	2,913,443	2,676,609
	Large	US$	(24,420,000)	7,874,692	7,874,692	7,874,692	7,322,228	6,797,386	6,298,787	5,825,118

NPV

			7 years
Small	US$m		(4.5)
Medium	US$m		2.3
Large	US$m		9.9

IRR

		7 years	5 years	10 years
Small		−28%	−37%	−27%
Medium		16%	9%	19%
Large		23%	17%	27%

Source: World Bank.

Note: Base case project duration from capex completion is modeled at 7 years. IRR sensitivity is modeled at 5 and 10 years from capex completion. AG = associated gas; capex = capital expenditures; CNG = compressed natural gas; IRR = internal rate of return; LPG = liquefied petroleum gas; mmscf/d = million standard cubic feet per day; NPV = net present value; opex = operating expenditures.

TABLE B.15 Liquefied natural gas to gas processing plant base case assumptions

MODEL VARIABLE	FLARE SIZE	UNIT OF MEASURE	VALUE	COMMENTS
Capex (liquefaction, trucks, gasification)	Small	US$/mmscf/d	10,000,000	
	Medium	US$/mmscf/d	6,200,000	
	Large	US$/mmscf/d	5,750,000	
Opex		% total capex	8%	Including for trucks, assuming they do not travel beyond a 150 km radius
Fuel gas ratio			98%	
Parasitic gas usage			15%	12% for liquefaction, 3% for dewpointing
LNG price		US$/mscf	8.00	
AG purchase price		US$/mscf	0.25	
Plant availability		% days in year	98%	

Source: World Bank.

Note: Base case project duration from capex completion is modeled at 7 years. AG = associated gas; capex = capital expenditures; km = kilometer; LNG = liquefied natural gas; mmscf/d = million standard cubic feet per day; mscf = thousand standard cubic feet; opex = operating expenditures.

TABLE B.16 Liquefied natural gas to gas processing plant base case calculations

PARAMETER	SIZE FLARE	UNIT	0	1	2	3	4	5	6	7
Gas flared (routine)	Small	mmscf/d	0.850	0.808	0.767	0.729	0.692	0.658	0.625	0.594
	Medium	mmscf/d	4.250	4.038	3.836	3.644	3.462	3.289	3.124	2.968
	Large	mmscf/d	8.500	8.075	7.671	7.288	6.923	6.577	6.248	5.936
Gas liquefied (per day)	Small	mmscf/d	0.833	0.791	0.752	0.714	0.678	0.645	0.612	0.582
	Medium	mmscf/d	4.165	3.957	3.759	3.571	3.392	3.223	3.062	2.909
	Large	mmscf/d	8.330	7.914	7.518	7.142	6.785	6.446	6.123	5.817
Capex	Small	US$	8,000,000							
	Medium	US$	22,320,000							
	Large	US$	41,400,000							
Revenues	Small	US$		1,737,175	1,737,175	1,737,175	1,650,317	1,567,801	1,489,411	1,414,940
	Medium	US$		8,685,877	8,685,877	8,685,877	8,251,583	7,839,004	7,447,054	7,074,701
	Large	US$		17,371,754	17,371,754	17,371,754	16,503,166	15,678,008	14,894,108	14,149,402
Opex	Small	US$		640,000	640,000	640,000	640,000	640,000	640,000	640,000
	Medium	US$		1,785,600	1,785,600	1,785,600	1,785,600	1,785,600	1,785,600	1,785,600
	Large	US$		3,312,000	3,312,000	3,312,000	3,312,000	3,312,000	3,312,000	3,312,000
Purchase of AG	Small	US$		63,867	63,867	63,867	60,673	57,640	54,758	52,020
	Medium	US$		319,334	319,334	319,334	303,367	288,199	273,789	260,099
	Large	US$		638,667	638,667	638,667	606,734	576,397	547,577	520,199
Capex to depreciate	Small	US$	8,000,000							
	Medium	US$	22,320,000							
	Large	US$	41,400,000							
Depreciation deflated by inflation	Small	US$		1,600,000	1,600,000	1,600,000	1,600,000	1,600,000	0	0
	Medium	US$		4,464,000	4,464,000	4,464,000	4,464,000	4,464,000	0	0
	Large	US$		8,280,000	8,280,000	8,280,000	8,280,000	8,280,000	0	0
Corporate tax base	Small	US$		(566,691)	(566,691)	(566,691)	(650,357)	(729,839)	794,653	722,920
	Medium	US$		2,116,943	2,116,943	2,116,943	1,698,616	1,301,205	5,387,665	5,029,002
	Large	US$		5,141,087	5,141,087	5,141,087	4,304,432	3,509,611	11,034,530	10,317,204

(continued)

TABLE B.16, *continued*

PARAMETER	SIZE FLARE	UNIT	YEAR							
			0	1	2	3	4	5	6	7
Corporate tax with loss carry-forward and no other taxable income	Small taxable income			(566,691)	(1,133,383)	(1,700,074)	(2,350,431)	(3,080,270)	(2,285,617)	(1,562,696)
	Small tax paid			0	0	0	0	0	0	0
	Medium taxable income			2,116,943	4,233,887	6,350,830	8,049,446	9,350,651	14,738,316	19,767,318
	Medium tax paid			0	0	0	0	0	0	0
	Large taxable income			5,141,087	10,282,173	15,423,260	19,727,692	23,237,302	34,271,833	44,589,036
	Large tax paid		0	0	0	0	0	0	0	0
Net cashflows	Small	US$	(8,000,000)	1,033,309	1,033,309	1,033,309	949,643	870,161	794,653	722,920
	Medium	US$	(22,320,000)	6,580,943	6,580,943	6,580,943	6,162,616	5,765,205	5,387,665	5,029,002
	Large	US$	(41,400,000)	13,421,087	13,421,087	13,421,087	12,584,432	11,789,611	11,034,530	10,317,204

NPV

		7 years
Small	US$m	(3.1)
Medium	US$m	6.8
Large	US$m	17.7

IRR

	7 years	5 years	10 years
Small	−5%	−15%	1%
Medium	20%	13%	24%
Large	24%	17%	27%

Source: World Bank.

Note: Base case project duration from capex completion is modeled at 7 years. IRR sensitivity is modeled at 5 and 10 years from capex completion. AG = associated gas; capex = capital expenditures; IRR = internal rate of return; mmscf/d = million standard cubic feet per day; NPV = net present value; opex = operating expenditures.

www.ingramcontent.com/pod-product-compliance
Lightning Source LLC
Chambersburg PA
CBHW041420290326
41932CB00042B/32